Compact Textbooks in Mathematics

This textbook series presents concise introductions to current topics in mathematics and mainly addresses advanced undergraduates and master students. The concept is to offer small books covering subject matter equivalent to 2- or 3-hour lectures or seminars which are also suitable for self-study. The books provide students and teachers with new perspectives and novel approaches. They may feature examples and exercises to illustrate key concepts and applications of the theoretical contents. The series also includes textbooks specifically speaking to the needs of students from other disciplines such as physics, computer science, engineering, life sciences, finance.

- **compact:** small books presenting the relevant knowledge
- **learning made easy:** examples and exercises illustrate the application of the contents
- **useful for lecturers:** each title can serve as basis and guideline for a semester course/lecture/seminar of 2-3 hours per week.

Howard Karloff

Mathematical Thinking

Why Everyone Should Study Math

 Birkhäuser

Howard Karloff
New York, NY, USA

ISSN 2296-4568   ISSN 2296-455X (electronic)
Compact Textbooks in Mathematics
ISBN 978-3-031-33202-9   ISBN 978-3-031-33203-6 (eBook)
https://doi.org/10.1007/978-3-031-33203-6

This book is published under the imprint Birkhäuser, www.birkhauser-science.com by the registered
company Springer Nature Switzerland AG
The registered company address is: Gewerbestrasse 11, 6330 Cham, Switzerland

Preface

This book is about math: fun math, the intriguing, fascinating math that (some) people love to do. This book is not about math you're forced to do. It's about the interesting kind you might not be getting enough of, or not have gotten enough of, in school.

Years ago I attended an Independence Day celebration at the vacation home of a friend. A friend of his young daughter had heard that I had majored in math in college. My friend's daughter and her friend approached me and the friend's daughter asked, "Why did you major in math in college?" Without thinking, I replied, "Why would you major in anything else?"

I will try to convince you that math is, if not the most interesting subject, then one of the most interesting ones. Some of you may be surprised to hear this, especially if you don't (or didn't) find math interesting in school. I contend that the math you learn in school is too focused on practical applications –what you can do with it –and not enough on the beauty of math itself. For example, everyone needs to learn the methods –known as the "algorithms"– for adding, subtracting, multiplying, and dividing integers. There's no doubt the algorithms are important. They're just not beautiful.

Practical uses, so-called *applications* of mathematics, are not the primary focus of this book, though we will spend some time on them. The focus of this book will be on the beauty and fun of math. It will be (mostly) about math *just for math*. I will try to convince you that math is like art, something elegant and beautiful, something you study because it is fun and beautiful.

I will try to show you how mathematicians really think, and to that end, I will be somewhat ambitious. There's a dichotomy in the mathematical book world between the popular books, each written in a casual tone, which try to excite their readers about math, while omitting the proofs, and the more formal textbooks, which include proofs and get all the details right. (If you don't know what a proof is, don't panic. You'll see numerous examples in this book.) I've tried to merge the genres here. I've tried to keep the tone casual while still including a bunch of proofs (though not all) with almost all details right. After all, proofs are an integral part of mathematics. You don't know for certain that a statement is true until you've proven it. You don't really understand math until you understand proofs. If you find the proofs too onerous, skip them, and just take my word that the statements they're proving (the *theorems*) are correct.

Since I studied math as an undergraduate and my other passion, computer science, as a grad student, I have included some programming problems in the book as well. Think of these as dessert: once you know the math, it's fun to see it in action. Programming makes the math practical. Now I know I said the focus of this book is math, not applications, but since programming is fun, I decided to include some Python programming exercises. If you don't know Python, you can install Python, for free, on your computer –I recommend Anaconda (https://www.anaconda.com/products/distribution)– and then learn it (see https://www.python.org/about/gettingstarted/). Even skipping the Python altogether wouldn't be a disaster. You just wouldn't get the thrill of having implemented the algorithms yourself. You wouldn't get the thrill of computing the greatest common divisor of two 100-digit numbers yourself. You wouldn't be able to draw fractals yourself.

How much math do you already need to know to read this book? Everything in this book can be understood by a smart high school senior. What the book really requires is not a deep knowledge of math but a thirst to learn.

Each chapter of this book ends with a puzzle (which has nothing to do with the content of the chapter). Years ago when I was a professor of computer science at Georgia Tech, in some of my courses I ended each lecture with a puzzle. The puzzles were a big hit, so much so that some students suffered through the lectures just to hear the puzzles at the end. I've appropriated the best puzzles from those classes and attached them to the chapters.

This book is intended for self-study and for use as a textbook. Back when I was teaching the freshman-level Understanding and Constructing Proofs course in the College of Computing at Georgia Tech, I would have loved to teach from this book. This book would be appropriate for an Introduction to Higher Mathematics or Foundations of Mathematical Proof course. It would be a good supplemental text for an Introduction to Discrete Math course in a mathematics or computer science department and maybe even for an Introduction to Graph Theory course.

Enjoy.

New York City, NY, USA   Howard Karloff
January, 2023

Contents

Primes

1

1.1   Introduction

We will start with the prime numbers, a very good place to start. You have probably heard about primes.

I will use the term *integer* to mean an element of the set $\{\ldots, -3, -2, -1, 0, 1, 2, 3, \ldots\}$. A positive integer is a *divisor* of an integer n or *divides n* if it divides into n with zero remainder. Equivalently, when you divide n by the divisor you get a result (the *quotient*) which is an integer. For example, 2 is a divisor of 6 because the remainder when dividing 6 by 2 is 0. Equivalently, $6/2 = 3$, which is an integer. By contrast, 2 is not a divisor of 7 because the remainder when dividing 7 by 2 is 1, and $7/2 = 3.5$, which is not an integer.

Now there are two kinds of divisors, the trivial ones and the nontrivial ones. The *trivial divisors* of a positive integer are 1 and the integer itself. We call them "trivial," which means "unimportant" or "of very little value," because these are divisors of every positive integer and they tell us nothing about the integer. For example, the trivial divisors of 12 are 1 and 12. The remaining divisors, if any, are the *nontrivial* divisors. They are the interesting ones. For example, the nontrivial divisors of 12 are 2, 3, 4, and 6. Now we get to the important definitions. A *prime* is an integer which is at least 2 and which has no nontrivial divisors. The first few primes are 2, 3, 5, 7, 11, 13, 17. All the other integers which are at least 2 are known as *composites*; these are the ones with at least one nontrivial divisor. The first few composites are 4, 6, 8, 9, 10, 12, 14, 15. The number 1 is neither prime nor composite.

An *even* integer n is one for which $n/2$ is an integer, in other words, one for which 2 divides n. The *odd* integers are all the others. All the primes, except 2, are odd. Can you see why? We will discuss this later.

© The Author(s), under exclusive license to Springer Nature Switzerland AG 2023
H. Karloff, *Mathematical Thinking*, Compact Textbooks in Mathematics,
https://doi.org/10.1007/978-3-031-33203-6_1

Let me whet your appetite with a question we will answer later: is the list of primes finite, meaning that it eventually stops, or infinite, so that it goes on forever?

1.2   Primes and Cryptography

Why are primes so important? Before I tell you why they're so important to math, I will tell you how practical they are. Do you use the Internet? Probably yes. Now some of the material you download or upload is not confidential, meaning that you wouldn't care if someone else read it. For example, you probably wouldn't mind if someone else saw the weather forecast you read. By contrast, some Internet communication is very confidential. When someone buys something on the Internet, the purchaser doesn't want anyone else to see the credit card or debit card number being used, since otherwise that *eavesdropper* (someone who can see everything being transmitted) could use that number to buy things. This would be theft and very illegal. For this reason, some communications on the Internet, such as those including credit card numbers, are "encrypted," which means that the message is disguised in such a way that only the person you want to see the message can understand it. Only a person with a secret key can decrypt the message. Without the secret key, the message is just gibberish and no more valuable than garbage.

Now not all Internet communication is encrypted. You know your communication is encrypted when you see a *locked* padlock next to an Internet address beginning with `https` (and not the usual `http`); in fact, the "s" stands for "secure."

Now here's the key point. In at least one popular encryption scheme (the "RSA Cryptosystem," invented by professors Rivest, Shamir, and Adleman), the secret keys are primes! Not primes like 7, but primes M like this one:

14759799152141802350848986227373817363120661453331697751477712164785

70297878078949377407337049389289382748507531496480477281264838760259

19181446336533026954049696120111343015690239609398909022625932693502

52814096149834993882228314485986018343185362309237726413902094902318

36446899608210795482963763094236630945410832793769905399982457186322

94472963641889062337217172374210563644036821845964963294853869690587

26504869144346374575072804418236768135178520993486608471725794084223

16678097670224011990280170474894487426924742108823536808485072502240

51945258754287534997655857267022963396257521263747789778550155264652

26099888699140135404838098656812504194976866697771007.

This is *one* long number, not a sequence of 10 numbers. It equals $2^{2203} - 1$, which is prime. (This is not obvious.)

I am not making this up. This *is* a prime. I would call this a "large prime."

(Decades ago, before the Internet, 4001 was considered a large prime. The meaning of "large prime" has really changed over the years, thanks to computers.)

This is one reason why primes are important. Primes are instrumental in enabling Internet commerce. People wouldn't purchase anything online if they thought their credit or debit card numbers would be stolen, and it's primes that, to some extent, make the Internet secure.

1.3   Back to the Math

From a mathematical perspective, primes are important because any positive integer can be written as a product of primes, possibly with repetitions. Let us call this fact the *Building Block Theorem*.

Theorem 1.1 *The Building Block Theorem. Every positive integer can be written as a product of zero or more primes.*

For example, $12 = 2 \cdot 2 \cdot 3$, $17 = 17$, and $825 = 3 \cdot 5 \cdot 5 \cdot 11$. The product of zero numbers is said to be 1, so that's how you write 1 as a product of primes.

We call this the *Building Block Theorem* as it shows how to use primes as building blocks to get any positive integer. A *theorem*, by the way, is a statement which has been proven to be true.

In fact, even more is true: there is only *one* way to represent a positive integer as a product of primes, if the primes have to be given in nondecreasing order. ("Nondecreasing order" means the numbers always increase or stay the same.) For example, in the factorization of 12, if you insist on writing the primes 2, 3 in that order, then $12 = 2 \cdot 2 \cdot 3$ is the *only* way to get 12. You can't use any primes except 2 and 3, and you must use exactly two 2's and one 3.

Now I want you to realize that something almost magical has happened here. We started with a simple definition of a prime as an integer at least 2 with no nontrivial divisors, and, a moment later, we stated the Building Block Theorem. *Where did it come from?* In math, and in much of life, you should be asking yourself, "Where did that come from?" and "Why is that true?" There really is something almost magical –not in the Disney sense of "magical," of course– in the fact that one can take such a simple definition and from it derive such an interesting theorem.

I am going to prove the Building Block Theorem to you. It is not hard, and it is such a foundational result; it's worth proving. To do it, we need two simpler theorems.

Theorem 1.2 *The Divisor-of-Divisor Theorem. If you have three positive integers written down, the first integer being a divisor of the second, and the second being a divisor of the third, then the first is a divisor of the third.*

If you want to be fancy and rewrite this in letters, it would read, "If a, b, c are positive integers, a a divisor of b and b a divisor of c, then a is a divisor of c."

Let's take an example before we tackle proving it. Consider 2, 6, 30. The second integer, 6, divided by the first integer, 2, gives a quotient of 3, which is an integer, so 2 is a divisor of 6. The third integer, 30, divided by the second integer, 6, gives a quotient of 5, an integer, so 6 is a divisor of 30. By the theorem, 2 should be a divisor of 30... which it is, because $30/2 = 15$, which is an integer. Notice that the quotient, which is 15, is the product of the two quotients, 3 and 5, that we got before. That is definitely not a coincidence.

Now let's prove the theorem.

Proof Let's use a, b, and c for the first integer, second integer, and third integer, respectively. We are told that a is a divisor of b and b is a divisor of c. Let $d = b/a$, and $e = c/b$. These quotients always exist. The interesting fact is that, since a divides b, $d = b/a$ is an *integer*, and since b divides c, $e = c/b$ is an *integer*. Now you know that $\frac{c}{a} = \frac{b}{a} \cdot \frac{c}{b} = d \cdot e$. But the product of two integers is an integer, so $d \cdot e$ is an integer. So we divided c by a and got an integer, which means that a is a divisor of c, and we are finished with the proof. ∎

We need one more theorem. This one is called the "Smallest Nontrivial Divisor Theorem," and it is quite simple.

Theorem 1.3 *The Smallest Nontrivial Divisor Theorem. The smallest nontrivial divisor of a composite number n is prime.*

Before embarking on proving this theorem, let's see a couple examples. Thirty is composite, and its smallest nontrivial divisor is 2, which is prime. 105 is composite, and its smallest nontrivial divisor is 3, which is prime. The smallest nontrivial divisor of 49, which is composite, is 7, which is prime.

Now we start the proof.

Proof Suppose n is composite. Now we use what it means for a number to be composite: n must have a nontrivial divisor. Since n has a nontrivial divisor, take, among all its nontrivial divisors, the smallest one. We don't know its identity, so let's call it x. Since x is nontrivial, $2 \le x < n$. Now x, being at least 2, is itself either prime or composite.

Now magic occurs: in fact, x must be prime! It cannot be composite. Why not? If x is composite, x itself has a nontrivial divisor, which we will call y. Now we have three integers y, x, n with $2 \le y < x < n$ with y a divisor of x and x a divisor of n. Now apply the Divisor-of-Divisor Theorem with a being y, b being x, and c being n. Since a is a divisor of b and b is a divisor of c, we know that a, which is y, is a divisor of c, which is n.

What does this mean? We have $2 \le y$, y is a divisor of n, and $y < x$. So y is a nontrivial divisor of n. But x was supposed to be n's *smallest* nontrivial divisor!

This is not possible, since y is a *smaller* nontrivial divisor of n. This means that x cannot be composite, so it must be prime, and hence the proof is complete.   ■

Finally, we are ready to prove the Building Block Theorem. We've done all the hard work already. We are going to use a technique called *induction*. This is a clever and very valuable way to prove something about all positive integers n. Here's how it goes.

1. Prove that the statement you want to prove is true for $n = 1$.
2. Prove, for *every* $n \geq 1$, that if the statement you want to prove is true for 1, 2, 3, ..., n, then it is also true for $n + 1$.

This probably looks complicated, but here's an informal argument as to why it works.

The first point above means the statement is true for 1.

The second point says (when $n = 1$) that if the statement is true for 1, which it is, then it's true for 2. Hence it *is* true for 2.

The second point says (when $n = 2$) that if the statement is true for 1 and 2, which it is, then it's true for 3. Hence it *is* true for 3.

The second point says (when $n = 3$) that if the statement is true for 1, 2, and 3, which it is, then it's true for 4. Hence it *is* true for 4.

The second point says (when $n = 4$) that if it's true for 1, 2, 3, and 4, which it is, then it's true for 5. Hence it *is* true for 5.

You see the pattern. In this way we prove the theorem for *every* positive integer n.

Formally, you prove that induction works by saying that if there is any n for which the statement is false, that is, a *counterexample*, then let n_0 be the smallest n for which the statement fails, that is, the smallest counterexample. (It is important here that if there are *any* counterexamples, then there actually is a *smallest* counterexample.) Since we checked that the statement holds for $n = 1$, we know that $n_0 \geq 2$. Therefore the statement holds for $n = 1, 2, 3, \ldots, n_0 - 1$. But the inductive assertion is that if the statement holds for all n's up to a particular value (in this case $n_0 - 1$), then it holds for the next value. So $n = n_0$ cannot be a counterexample.

Now we prove the Building Block Theorem by induction.

Proof We follow the inductive framework.

First we prove the statement for 1. By definition, 1 is the product of zero primes, so the statement is true for 1.

Now take any $n \geq 1$, and show that if every positive integer between 1 and n is the product of zero or more primes, then $n + 1$ is also the product of zero or more primes. If $n + 1$ is prime, then of course $n + 1$ is the product of zero or more primes. If not, then $n + 1$ is composite, and it has a smallest nontrivial divisor, which has to be a prime, so we'll call it p. Let $k = (n + 1)/p$. Hence k is an integer which is less than $n + 1$ and hence at most n. By the inductive hypothesis, because $k \leq n$, we know that k is itself the product of zero or more primes. That means there are primes, say, $q_1, q_2, \ldots, q_h$, such that

$$k = q_1 \cdot q_2 \cdot q_3 \cdots q_h.$$

But

$$n + 1 = p \cdot k = p \cdot q_1 \cdot q_2 \cdot q_3 \cdots q_h,$$

which is a (slightly longer) product of primes.

Because we have proven both parts required for an inductive proof, the proof is complete.   ∎

Let's return now to a simple claim made earlier, that all primes except for 2 are odd.

Theorem 1.4 *All primes except for 2 are odd.*

You probably know how to prove this already.

Proof Let n be any even integer which is at least 4. We will show that n cannot be prime, so the only primes bigger than 2 must be odd. (Notice that we are not showing that all odd positive integers at least 4 are prime, which is false; we are showing that even positive integers at least 4 cannot be prime.)

Since n is even, $n/2$ is an integer, which means that 2 is a divisor of n. But is 2 a *nontrivial* divisor of n? Clearly $1 < 2$. Because $4 \le n$, $2 < n$. Hence $1 < 2 < n$, making 2 a nontrivial divisor of n, and hence n is composite and not prime.   ∎

By the way, in math circles there is an old joke about 2. As the only even prime, 2 is an extremely unusual prime. That makes 2 an odd prime, indeed.

Now we get to a really fun result alluded to earlier. How many primes are there, finitely many (so that the list of primes stops at some point), or infinitely many (so that the list goes on forever)? You probably guessed that the list is infinite. It is. Here's a simple but beautiful proof by contradiction, but let me say a few words about "proofs by contradiction."

In a proof by contradiction, one adopts the belief that the statement one wants to prove is *false* (even though we believe it's not). Then we show a *contradiction*— something which is impossible, like $2 = 3$ or 6 is prime. Since contradictions are not possible, it must be the case that the statement we want to prove cannot be false and hence must be true.

In the next proof, since we want to show that the list of primes is infinite, we start by pretending that the list of primes is finite, and we derive a contradiction.

Theorem 1.5 *The list of primes is infinite.*

Proof Suppose the list of primes is finite. Then because this list is finite, we can multiply together all the primes on this list –you couldn't do this if the list of primes were infinite– calling the product, say, z. Now z is a multiple of all the primes in the world (because the list contained all the primes in the world). Now consider $z + 1$. Because it is one greater than z, it is the opposite of z, in that it is a multiple of *none* of the primes in the world.

The Building Block Theorem says every positive integer is the product of zero or more primes. However, $z + 1$, which is at least 3, cannot be written as a product of zero primes. So there must be a prime dividing $z + 1$, but $z + 1$ is a multiple of *no* prime, which is impossible. We have shown that the list of primes is infinite. ■

This result is so important, and this is so much fun, that I'll give a second proof. The previous proof was by contradiction, whereas this one –which is very similar to the previous one– is *constructive*, meaning that it shows you how to construct infinitely many primes.

Proof To generate an infinite list of primes, it is enough to show, given any finite list of primes, how to generate a new prime, for one can apply such a procedure repeatedly forever.

Suppose you have a finite list L of primes. Let z be the product of all the primes on the list. Consider $z + 1$. When $z + 1$ is divided by any prime on the list L, the remainder is 1, and hence none of those primes is a divisor of $z + 1$. If $z + 1$ is prime itself, it's a prime which is not on the list. If $z + 1$ is not prime itself, Theorem 1.3 shows that $z + 1$ has a divisor, say, p, which is prime (and not on the list). In either case we have a prime which is not on the list, so we have extended the list of primes by one. ■

1.4   Algorithms for Testing Primality

If I gave you a positive integer n, how could you tell if it's prime or not? That's the problem of determining *primality*. Remember, a prime is an integer which is at least 2 and has no nontrivial divisors. The obvious way to check if n is prime is to try all potential nontrivial divisors, from 2 up to $n - 1$, one at a time. That is, see if 2 divides n; see if 3 divides n; see if 4 divides n; . . . , see if $n - 1$ divides n. The number n is prime if all of those answers are "no." If you know how to program a computer at all, write yourself a little program to test if n is prime—I recommend downloading Anaconda Python, which is free, and coding in Python. You can use the Python program `primality.py` on the book's website if you don't want to code it yourself.

This is good, but with one little improvement, one can speed up the primality test. Do we really need to try all potential nontrivial divisors? Suppose we're testing if 127 is prime, and suppose we've tested $2, 3, 4, \ldots, 11$ already, and discovered that none of them divides 127.

Do we need to check 12? Now $12 \cdot 12 = 144$, so the quotient $127/12 < 144/12 = 12$. This means the quotient, which we'll call q, if an integer, is at most 11 and *must also be a divisor of 127*. But we're already tried $2, 3, 4, 5, 6, 7, 8, 9, 10, 11$, and none were divisors of 127... which means that 12 cannot be a divisor of 127.

Do we need to check 13? The quotient $127/13$ is less than the quotient $127/12$ we just discussed, so if it's an integer, it must divide 127 and must be between 2 and 11. However, there are no divisors of 127 between 2 and 11. Hence 13 cannot be a divisor of 127, either.

By the same reasoning, there's no need to check $14, 15, 16, \ldots, 126$.

What have we learned? The key point is that we can stop checking trial divisors d when $d \cdot d > n$. Program `fasterprimality.py`, also on the website, incorporates this improvement.

By the way, checking if that huge number M above is prime requires much, much more sophisticated methods. Even `fasterprimality.py` would be way too slow. These very sophisticated methods *do* exist; one *can* check if a 1000-digit number is prime.

1.5   *e* and the Prime Number Theorem

1.5.1   *e*

The number e is one of the most important numbers in mathematics. It is just a real number, but a superimportant one. Its value is approximately 2.718281828459. Maybe e is a little scary... but it shouldn't be. It's just, like π, a really interesting real number!

The best way I know of to define e is through banking. Suppose you put $1 into Bank 1 for a year into an account paying 100% interest.[1] At the end of the year, the bank pays you $1 additional, which it credits to your account, and you then have $2 in your account. Simple! Seriously, this is called *simple interest* or *annual compounding*. Since the interest rate r is 100%, or 1.0, at the end of the year, they multiply your initial balance of $1 by $1 + r = 2$.

However, Bank 2 offers you a better deal: they offer to compound the interest twice a year. This means that after half a year, they pay you $0.50, adjusting your balance to $1.50. In the second half of the year, you get interest on your original $1, and *interest on the interest*, that is, they pay you 50% interest (since it's for only half a year) on $1.50, or $0.75. At the end of the year, you have $2.25! Another way to see this is that Bank 2 is multiplying your balance by $1 + r/2 = 1.5$, but they're doing it *twice* ($1.5^2 = 2.25$). Bank 2 is giving you a better deal than Bank 1.

---

[1] Good luck finding a bank paying 100% interest.

Yet now Bank 3 steps in, and offers to compound your interest three times a year. This means they're multiplying your balance by $1 + r/3 = 1.333333\ldots$, but they do it three times a year. Since $r = 1$, at the end of the year, you'll have a balance of $\$(1 + r/3)^3 = \$(4/3)^3 \approx \$2.37$. ("$\approx$" means "is approximately equal to.") This is more than you had at Bank 2.

There's no reason, at least in principle, to stop with thrice-yearly compounding. You can certainly imagine Banks $4, 5, 6, 7, \ldots$ and so on. Each is more generous than the previous one.

What happens if you keep going *ad infinitum*? You have greater and greater bank balances. As you look at Bank n for larger and larger n, do your limiting bank balances go to infinity or do they approach a limit? The answer is that they approach a finite limit –this is called *continuous compounding*– and that limit is e, which, as I wrote, is approximately 2.718281828459. This is the number of dollars you'll have in the bank after 1 year if you deposit \$1 at the beginning of the year and you receive 100% interest compounded continuously.

I can't overemphasize the importance of e in mathematics and science. Did you ever study logarithms? The logarithm of a positive number, to the base 10, is the number of tens you have to multiply to get that positive number. For example, the logarithm (or *log*) of 10 to the base 10 is 1, since $10 = 10^1$. The log of 100 to the base 10 is 2, since $100 = 10^2$. Since 50 is more than 10 and less than 100, its log to the base 10 is between 1 and 2; in fact, it's approximately 1.699.

However, there's no good reason to use 10 as the base of the logarithm. The only reason I can think of is that we use a decimal number system. It turns out that a much more natural base of a logarithm is e. In fact, it's so natural to use e as the base, the log defined this way is called a *natural logarithm*, denoted $\ln(x)$. Though I haven't examined every scientific calculator ever made, I'm not going out far on a limb to state that every scientific calculator ever made has a button for $\ln(x)$; that's how important natural log is. The natural log of a number x is just the number y such that $e^y = x$. For example, the natural log of e is 1, since $e = e^1$ (of course). The natural log of 10 is approximately 2.3, since $10 \approx e^{2.3}$.

The function $f(x) = e^x$ has a remarkable property you probably saw if you studied calculus. If one computes the slope (also known as "derivative") of the function $f(x) = e^x$ at the point (x, e^x), the slope is e^x itself! That is, the slope of the function equals the value of the function, for every x. It is certainly not obvious that this is true. We'll see more about slopes in Chap. 8 on Newton's method.

1.5.2   The Prime Number Theorem

Wouldn't it be nice to have a simple formula for the nth prime number? Wouldn't it be nice if one could plug in 117,423 into the formula and have it spit out the 117,423rd prime number? There certainly is a method to take in n and output the nth prime number: just start from 2, crossing out composites on the way, until you have accumulated n prime numbers. Unfortunately, this is an algorithm, but no one would call it a simple formula. In fact, no one has ever found a simple formula for the nth prime number, and there are good reasons to suggest none exists.

Instead, people have made progress on a similar question: given a positive integer n, how many prime numbers are less than or equal to n? Let the number of primes at most n be called $\pi(n)$. (I apologize for the confusion. This π is the same letter $\pi \approx 3.14$ that you know and love, but $\pi(n)$ has *nothing* to do with that π.) For example, the primes at most 10 are 2, 3, 5, and 7, so $\pi(10) = 4$. The primes at most 20 are 2, 3, 5, 7, 11, 13, 17, and 19, so $\pi(20) = 8$.

Now the fun begins. Clearly $\pi(n) \le n$, so $\pi(n)/n \le 1$, but by how much is $\pi(n)$ smaller than n? It's not hard to see that the primes "thin out" as you look at larger and larger numbers. For example, there are more primes between 1 and one million than between one million and two million. This means that you wouldn't expect $\pi(n)/n$ to be approximately a *positive* constant. Here's the remarkable result.

Theorem 1.6 *The Prime Number Theorem.* $\frac{\pi(n)}{n/\ln(n)}$ *approaches 1 as n approaches infinity.*

This means that $\pi(n)/n$, the fraction of primes between 1 and n, is approximately $1/\ln(n)$, when n is large. For example, $\pi(10^6) = 78,498$, whereas $10^6/\ln(10^6) = 72,382.4\ldots$, so $\pi(n)/(n/\ln(n))$ for $n = 10^6$ is about 1.08449. Since $\pi(10^9) = 50,847,534$ and $10^9/\ln(10^9) = 48,254,942.4\ldots$, the ratio is approximately 1.05373. Last, $\pi(10^{12}) = 37,607,912,018$, and since $10^{12}/\ln(10^{12}) = 36,191,206.8\ldots$, $\pi(n)/(n/\ln(n))$ for $n = 10^{12}$ is approximately 1.039.

Let's pause for a moment to appreciate how astounding this theorem is. Who would have guessed that the density of primes between 1 and n approaches one divided by the logarithm of n, the logarithm to the base equal to that number e defined above via compound interest? Why is there a connection between e and primes, anyway?

Unfortunately, the proof of the prime number theorem is too complex to include in this book.

1.6 Two Fascinating Questions

1.6.1 Goldbach's Conjecture

Let's do some calculating.

4 is even and is the sum of two primes: $4 = 2 + 2$.
6 is even and is the sum of two primes: $6 = 3 + 3$.
8 is even and is the sum of two primes: $8 = 3 + 5$.
10 is even and is the sum of two primes in two different ways: $10 = 3+7 = 5+5$.
12 is even and the sum of two primes: $12 = 5 + 7$.
14 is even and is the sum of two primes: $14 = 3 + 11 = 7 + 7$.
16 is even and is the sum of two primes: $16 = 3 + 13 = 5 + 11$.
18 is even and is the sum of two primes: $18 = 5 + 13 = 7 + 11$.
20 is even and is the sum of two primes: $20 = 3 + 17 = 7 + 13$.

You see the pattern, of course. After examining this pattern, you are likely to make the following conjecture. A *conjecture* is a mathematical statement which you believe to be true but which hasn't yet been proven.

Conjecture 1.1 Goldbach's Conjecture. Every even integer at least 4 is the sum of two primes.

This is a remarkably easy-to-understand conjecture. If you've seen primes, and many kids see them in elementary school, then you understand the conjecture.

Is it true or not? Remarkably, as of early 2023, *no one knows*. All of the mathematicians in the world, and all now-dead mathematicians who lived in the past, could not prove or disprove this conjecture.

Proving the conjecture would somehow involve showing that every $n \geq 4$ is the sum of two primes. In other words, that the list of even $n \geq 4$ which can be written as a sum of two primes is exactly this infinite list: 4, 6, 8, 10, 12, 14, 16, ... Now sometimes, showing that a list goes on forever isn't that hard. Showing that the list of primes goes on forever, as we saw above, isn't so hard. In this case, it's superhard.

To *disprove* a conjecture means to prove that it is false. In the case of Goldbach's conjecture, one could disprove it simply by giving one positive integer $n \geq 4$ and showing, by listing the possibilities, that n cannot be written as the sum of two primes. Such an n would be a *counterexample*, which means "an example which proves the conjecture false." This sounds easy. . . if there *is* such an n. If there isn't one, then of course you won't find one. Of course people have tried to find such an n, but no one's ever found one. It is known that the smallest counterexample n must be at least $4 \cdot 10^{18}$, so you can try any $n < 4 \cdot 10^{18}$ and you won't find a counterexample.

1.6.2   Twin Primes

Now that we know what a prime is, can you guess what a *twin prime* is? A *twin prime* is a prime such that the number two larger is also prime, for example, 3 (because 5 is prime), 5 (because 7 is prime), 11 (because 13 is prime), 17 (because 19 is prime), etc.

By the way, you might be asking yourself if there's a concept of a *triple prime*, three prime numbers, a prime, like 3, so that the numbers 2 and 4 greater are both prime. Yes, there is –3,5,7 of course– but that's the only one. Can you prove that yourself?

Let's ask a question similar to the one we asked earlier about the list of primes. Is the list of twin primes finite (so that it stops at some point) or infinite (so that it goes on forever)? Remarkably, as in the case of Goldbach's conjecture, no one knows. The best mathematicians alive today or from the past have been unable to prove or disprove this conjecture.

Conjecture 1.2 The Twin Prime Conjecture. The list of twin primes is infinite.

We showed already that the list of primes is infinite, but no one's been able to do the same for the list of twin primes. Maybe *you* will some day.

1.7   Puzzle

You are being held in a windowless jail cell in an autocratic country. There are three light switches in your cell, one of which controls an incandescent light bulb in the adjacent cell. You have 60 minutes to figure out which light switch controls the bulb. After 60 minutes, you may leave the cell to examine the bulbs, but once you do so, you cannot touch the switches again. If you solve the puzzle, you will go free; if you fail, you will be executed immediately. What do you do?

1.8   Exercises

Exercise 1.1 Using the method of trial division up to $\sqrt{n}$ to test if a number n is prime, on a computer doing a billion trial divisions per second, how long would it take to determine if a given 1000-digit number is prime?

Exercise 1.2 We know that when testing n for primality, it suffices to try trial divisors between 2 and $\sqrt{n}$. Show that it suffices to try *prime* trial divisors between 2 and $\sqrt{n}$ (assuming you have a list of the primes between 2 and $\sqrt{n}$).

Exercise 1.3 If a, b are nonnegative integers, not both 0, their *greatest common divisor* is the largest integer g dividing both a and b. For example, the greatest common divisor of 9 and 30 is 3. Two nonnegative integers, not both 0, are *relatively prime* if their greatest common divisor is 1.

(a) Give a pair of relatively prime positive integers neither of which is prime.
(b) Show how to tell if two integers are relatively prime given their prime factorizations.

Unfortunately, no one knows a fast algorithm for finding prime factorizations of large integers, say, those with 1,000 digits. In Chap. 2, we will see a very fast algorithm for testing if two numbers are relatively prime *without finding their prime factorizations*. It can even work on 1,000-digit numbers.

Exercise 1.4 According to Wikipedia as I write this, the general number field sieve is the fastest known algorithm for factoring integers greater than 10^{100}. Let n be a very large integer which is hard to factor. The running time of the general number field sieve to factor an integer n is approximately

$$e^{[(64/9)^{1/3}(\ln n)^{1/3}(\ln(\ln n))^{2/3}]}.$$

(This is the worst running time. It is not the time to factor easy 1000-digit numbers, like 10^{999}, but instead to factor hard numbers to factor, like those that are products of two roughly 500-digit primes.)

Assuming one billion operations per second, how much time would it take to factor one hard-to-factor 1000-digit number, roughly?

Exercise 1.5 * Prove that the only n such that $n, n+2, n+4$ are all prime is $n = 3$.

Exercise 1.6 * Show that for any positive integer k, there are k consecutive positive integers none of which is prime.

Exercise 1.7 * The constructive proof of the infinitude of primes shows that if S is any finite set $\{q_0, q_1, q_2, \ldots, q_i\}$ of primes of product z, then there is a prime $q_{i+1} \leq 1 + z$ which is not in S. Start with $q_0 = 2$ and $S = \{q_0\}$. Using the proof, generate an infinite list of primes $< q_0, q_1, q_2, \ldots >$.

(a) Prove that $q_i \leq 2^{2^i}$ for all i.
(b) Prove that $\pi(n) \geq \log_2(\log_2(n))$ for all $n \geq 2$. Here $\log_2(x)$ is the real y such that $2^y = x$. Hence, for example, $\log_2(\log_2(18)) \approx \log_2(4.1699) \approx 2.06002$.

1.8.1 Hints

1.5 Look for a multiple of 3 among $n, n + 2, n + 4$.
1.6 Think about $k!$.
1.7 Weaken the condition to $q_{i+1} \leq 2z$ and prove $q_i \leq 2^{2^i}$ by induction.

The Euclidean Algorithm

2

2.1 The Extended Euclidean Algorithm

In this chapter we will first study a simple algorithm, based on elementary-school division, to compute greatest common divisors. This algorithm (i.e., method) was discovered by ancient Greek mathematician Euclid. An extension of this algorithm will allow us to prove uniqueness of prime factorization, the main result of this chapter.

You have been hearing quite a bit about the ancient Greeks. Don't feel bad if you don't remember the names. Names are unimportant in mathematics. You should remember and understand the concepts. What they're called is unimportant.

Here's a new definition.

Definition 2.1 If m and a are integers and $m > 0$, define $a \bmod m$ to be the unique remainder r between 0 and $m - 1$ when a is divided by m. That is, r is the unique integer in $\{0, 1, 2, \ldots, m - 1\}$ such that there is an integer q such that $a = qm + r$.

Hence $7 \bmod 4 = 3$ and $(-7) \bmod 4 = 1$.

I'm sure you've seen this next definition.

Definition 2.2 Where a, b are integers, $a|b$ means that a is a divisor of b, i.e., that there is an integer k such that $b = ka$.

Definition 2.3 If $a, b \geq 0$ but are not both 0, the *greatest common divisor* $\gcd(a, b)$ is the largest integer g such that $g|a$ and $g|b$.

H. Karloff, *Mathematical Thinking*, Compact Textbooks in Mathematics, https://doi.org/10.1007/978-3-031-33203-6_2

You've used gcd's when reducing fractions to lowest terms. The number by which to divide the numerator and denominator when reducing a fraction to lowest terms is the gcd of the numerator and denominator. For example, the gcd of 400 and 700 is 100, so $\frac{400}{700}$ in lowest terms is $\frac{400/100}{700/100} = \frac{4}{7}$.

Now we need an important theorem that will suggest the path to a fast method to compute $\gcd(a, b)$.

Theorem 2.1 *If* $0 < a \le b$, *then* $\gcd(a, b) = \gcd(b \bmod a, a)$.

Proof Let C be the set of all common divisors of a and b (i.e., positive integers which divide both a and b), and let C' be the set of all common divisors of $b \bmod a$ and a (i.e., positive integers which divide both $b \bmod a$ and a). We will show that $C = C'$, and hence $\gcd(a, b)$, which is the largest element in C, is the same as the largest element in C', which is $\gcd(b \bmod a, a)$.

To prove that set C equals set C', we will show that every element of C is in C' and that every element of C' is in C.

First we show that every element of C is in C'. Let d be any element of C. That means that $d|a$, $d|b$. Now $b \bmod a = b - ka$ for some integer k. Since $d|b$, there is an integer q_1 such that $b = dq_1$. Since $d|a$, there is an integer q_2 such that $a = dq_2$. Hence $b - ka = dq_1 - kdq_2 = d(q_1 - kq_2)$, so $d|(b - ka)$. Since $d|a$, d divides both $b \bmod a$ and a, so d is in C'.

Now let's show that every element of C' is in C. Let d be any element of C'. Hence $d|(b \bmod a)$ and $d|a$. We have $b \bmod a = b - ka$ for some integer k. But now $b = (b \bmod a) + ka$. Since $d|(b \bmod a)$ and $d|a$, it follows that $d|b$. So $d|b$, $d|a$. Hence d is in C.   ∎

How can we use this theorem in order to compute greatest common divisors? It's simple. Take a, b with $0 \le a \le b$, $b > 0$. We will repeatedly replace the pair of integers whose gcd we want by a smaller pair with the same gcd.

Here's the Euclidean Algorithm for computing $\gcd(a, b)$:

1. If $a = 0$, return b as the gcd.
2. Now we have $0 < a \le b$. While $a > 0$, repeat:
   (a) Replace the pair (a, b) by the pair $(b \bmod a, a)$.
3. Return b as the gcd.

Example. Let's find the gcd of 105 and 426. It's certainly not obvious what it is. Let $a = 105$ and $b = 426$. As required, $0 \le a \le b$.

- In line 1, a is not 0, so don't return b as the gcd.
- We have $a = 105$ and $b = 426$. Now we execute line 2 repeatedly until a becomes 0:
  - Because 426 mod 105 = 6, replace a by 6 and b by 105.
  - Because 105 mod 6 = 3, replace a by 3 and b by 6.
  - Because 6 mod 3 = 0, replace a by 0 and b by 3.

- Finally, a is 0, so we stop repeating and return $b = 3$ as the gcd.

The next result is called *Bézout's Lemma* –"Bézout" is pronounced as "bay-zoo"– despite the fact that it was proven by Claude Gaspard Bachet de Méziriac. Usually results are named after the person who proved them. This naming of a result after the wrong person occurs often in mathematics. Only now, from Wikipedia, did I learn that the result I'm about to show you is called Bézout's Lemma. (A theorem used in proving something more important is called a *lemma*.) Remember that it's not the name that counts; it's the result.

Lemma 2.1 *Bézout's Lemma. Suppose $g = \gcd(a, b)$. Then there are integers r, s such that $r \cdot a + s \cdot b = g$.*

Why is this interesting? If r, s exist, since clearly $g|a$, we have $g|(ra)$. Similarly, $g|b$, and so $g|(sb)$. Hence $g|(ra+sb)$; that is, $ra+sb$ is always a multiple of g. The smallest positive multiple of g is g itself. So this lemma tells us that it is possible to get the *smallest* positive multiple of g, namely, g itself, by taking a combination $ra + sb$ of a and b. This is far from obvious.

In fact, something stronger holds. If h is *any* multiple of g, say, $h = lg$, then there are integers r', s' such that $r'a + s'b = h = lg$. This follows easily from Bézout's Lemma. Bézout's Lemma gives r, s so that $ra + sb = g$. Set $r' = lr$ and $s' = ls$; then $r'a + s'b = lg$.

For example, as we showed above, $\gcd(105, 426) = 3$. Why should there be integers r, s with $r \cdot 105 + s \cdot 426 = 3$? This is not obvious at all. In fact, now that I've told you that r, s exist, how would you find them?

Proof of Bézout's Lemma The proof involves a simple extension to the Euclidean algorithm, giving the not-so-creatively named *Extended Euclidean Algorithm*.

We start with an easy case. If $a = 0$ and $b > 0$, clearly $\gcd(a, b) = b$. Let $r = 0$ and $s = 1: ra + sb = b$.

So assume $a > 0$. Run the Euclidean Algorithm as before. After finishing, follow the steps *backward*. We will see how to take the pair (r, s) for one pair, and build a new pair (r', s') for the previous pair.

Let $g = \gcd(a, b)$. At the end $a = 0, b > 0$; $g = b$. We can take $r = 0, s = 1$, because $ra + sb = b = g$.

Now suppose we have r, s for the pair $(b \bmod a, a)$, that is, $r \cdot (b \bmod a) + s \cdot a = g$. We want to find r', s' for the previous pair (a, b), that is, $r'a + s'b = g$.

Recall that there is an integer k so that $(b \bmod a) = b - ka$. We have $r(b - ka) + sa = g$, i.e., $rb - rka + sa = g$. That is, $(s - rk)a + rb = g$. We want $r'a + s'b = g$. From $(s - rk)a + rb = g$, it is clear that $r' = s - rk, s' = r$ will work. ∎

Example. Suppose $a = 105, b = 426$. As we saw above, here are the pairs generated by the Euclidean Algorithm, and the associated values of k:

1. $(105, 426), k = 4$.

2. $(6, 105)$, $k = 17$.
3. $(3, 6)$, $k = 2$.
4. $(0, 3)$.

Now work backward. Here are the (r, s) pairs:

1. $(0, 1)$ for pair $(0, 3)$;
   $0 \cdot 0 + 1 \cdot 3 = 3 = \gcd(0, 3)$.
2. $(1 - 0 \cdot 2, 0) = (1, 0)$ for pair $(3, 6)$;
   $1 \cdot 3 + 0 \cdot 6 = 3 = \gcd(3, 6)$.
3. $(0 - 17 \cdot 1, 1) = (-17, 1)$ for pair $(6, 105)$;
   $-17 \cdot 6 + 1 \cdot 105 = 3 = \gcd(6, 105)$.
4. $(1 - 4 \cdot (-17), -17) = (69, -17)$ for pair $(105, 426)$;
   $69 \cdot 105 - 17 \cdot 426 = 3 = \gcd(105, 426)$.

It is an important point, by the way, that prime factorizations of a and b are unnecessary when running the Extended Euclidean Algorithm. The Extended Euclidean Algorithm always works. It has nothing to do with primality of a or b or their prime factorizations if they're not prime. In fact, if you have prime factorizations of a and b, then you can easily find $\gcd(a, b)$ (how?); the magic of the Euclidean Algorithm is that you can find gcd's *without* prime factorizations.

How fast is the (Extended) Euclidean Algorithm? Remarkably fast, in fact. Let's count the number of iterations through the *while* loop in the algorithm.

Lemma 2.2 *The value of b decreases by at least a factor of 2 after every pair of consecutive iterations.*

Proof The sequence of b's is strictly decreasing. In each iteration, the pair (a, b) is replaced by the pair $(b \bmod a, a)$. If $a \leq b/2$, then in one iteration the b value has dropped already by a factor of 2. If $a > b/2$, then the first iteration replaces (a, b) by $(b \bmod a, a)$, and the second one replaces that pair by a pair ending in $b \bmod a$. But if $a > b/2$, then $b \bmod a = b - a < b/2$.   ∎

Definition 2.4 The *binary logarithm* or *binary log* of a positive real x, written $\log_2(x)$, is the real y such that $2^y = x$.

Theorem 2.2 *The number of iterations of the Euclidean Algorithm and Extended Euclidean Algorithm when run on pair (a, b), $0 < a \leq b$, is at most $1 + 2\log_2 b$.*

Proof Divide the iterations of the algorithm into consecutive pairs with at most one iteration left over. Suppose those are m consecutive pairs. By the end of the m pairs of iterations, the value of b has been reduced, but it must be at least 1. Since each pair of iterations divides b by at least a factor of 2, we must have $b \geq 2^m$, and so $\log_2 b \geq m$. Since there can be at most one iteration after the m pairs of iterations, the total number of iterations is at most $1 + 2m \leq 1 + 2\log_2 b$.   ∎

Is this fast or slow? It's fast. The binary logarithm even of a 1000-digit number is about 3318, and 6637 is a small number of iterations for a modern computer to do.

Why does it matter? Not only are the Euclidean and Extended Euclidean Algorithms elegant, they are key ingredients in modern cryptography. In particular, the RSA encryption algorithm uses the Euclidean Algorithm.

You might be asking, as alluded to above, why not just find the prime factorizations of a and b in order to find $\gcd(a, b)$? That is a natural question, since given the prime factorizations of a and b, finding $\gcd(a, b)$ is trivial. That is an excellent and surprisingly interesting suggestion. It turns out that *no one knows a fast method to find prime factorizations for all large integers.* By "large integer," I don't mean 4001, I mean one with around 1000 digits. I don't mean that factoring *every* 1000-digit number into primes is difficult –it's easy to factor 1 followed by 999 zeros, which equals 10^{999}, into $2^{999} \cdot 5^{999}$– but that no one can factor *every* 1000-digit number quickly. The particularly difficult 1000-digit integers to factor are the products of two primes each with about 500 digits. Not only does no one know a fast method, it is believed by some to be impossible.

There's more. It is a fascinating fact that if you *could* find prime factorizations of any 1000-digit integer quickly, then you could break the RSA encryption scheme and decrypt all the RSA-encrypted messages flying across the Internet, get all the credit and debit card numbers, and steal all you wanted. That you could decrypt if you could factor is not obvious at all. Using today's algorithms, even the world's fastest computers would take a thousand years to factor a very large integer.

When I say, "no one knows a fast method to find prime factorizations for all large integers," I mean, "no one has publicly presented a fast algorithm to find prime factorizations for large integers." It is conceivable, for example, that someone or some agency, perhaps the US's National Security Agency, for example, already knows how to break the RSA encryption scheme. Personally, I doubt it, but this possibility cannot be ruled out.

2.2 Uniqueness of Factorization into Primes

In Chap. 1, we proved that any positive integer can be factored into prime numbers, and I stated, without proof, that the factorization is unique. Now you might object: $90 = 2 \cdot 3 \cdot 3 \cdot 5$ and $90 = 5 \cdot 3 \cdot 2 \cdot 3$, so why did I say the factorization is unique? What I meant was that the factorization is unique provided the prime divisors must be written in nondecreasing order. With this requirement, the only way 90 can be factored into primes is $90 = 2 \cdot 3 \cdot 3 \cdot 5$.

This claim of uniqueness of prime factorizations is really somewhat outlandish. After all, if you were factoring into arbitrary positive integers, each at least two, but *not* required to be primes, then uniqueness would be wildly false, e.g., $90 = 6 \cdot 15 = 9 \cdot 10$. What is it about primes that prevents this from happening? When you think about it, it's really not obvious at all. How do you know there isn't a very large integer n, say, one with 1000 digits, that can be factored into primes as $n = 2 \cdot 7 \cdot 13 \cdots$ and also into primes as $n = 3 \cdot 7 \cdot 11 \cdots$? This is really not obvious. Think about this point for a few minutes: how would *you* prove uniqueness of prime factorizations?

To prove uniqueness of prime factorizations, we will use the Extended Euclidean Algorithm. We will start by proving a natural lemma which was known to the ancient Greeks (or more precisely to at least one ancient Greek, Euclid, who was named after the Euclidean Algorithm[1]). However, first we need an important definition.

Definition 2.5 Fix a positive integer m. If a, b are integers (which may be negative), we say $a \equiv b$ (mod m), read "a is *congruent to b (mod m)*," if $a - b$ is a multiple of m.

For example, $100 \equiv 82$ (mod 9), because $100 - 82$, which is 18, is a multiple of 9, and $-3 \equiv 5$ (mod 4), because $-3 - 5$, which is -8, is a multiple of 4.

We will often use without mentioning it that for any m, $\equiv$ is an *equivalence relation*, that is,

- $\equiv$ is *reflexive*, i.e., $a \equiv a$ (mod m) for any a;
- $\equiv$ is *symmetric*, i.e., $a \equiv b$ (mod m) implies that $b \equiv a$ (mod m), for any a, b; and
- $\equiv$ is *transitive*, i.e., $a \equiv b$ (mod m) and $b \equiv c$ (mod m) imply that $a \equiv c$ (mod m) for any a, b, c.

In addition, if $a \equiv b$ (mod m), and x is integral, then $ax \equiv bx$ (mod m).

And now, a very important lemma.

Lemma 2.3 *Suppose that p is a prime. Suppose a and b are integers and $ab \equiv 0$ (mod p). Then either $a \equiv 0$ (mod p) or $b \equiv 0$ (mod p). Equivalently, if p divides ab, then either p divides a or p divides b.*

This doesn't seem obvious to me. It's not even true if p is not prime. For example, consider $p = 6$, $a = 10$, and $b = 9$; we have $ab = 90$, so $p|(ab)$, but p divides neither a nor b. So the lemma *must* require that p be prime, for otherwise it's false, but why is the lemma true when p *is* prime?

Let me call your attention to the analogue with the real numbers. If you have two nonzero *real* numbers a and b, then their product cannot be 0. The same sort of phenomenon is occurring here. In this sense, multiplying when using a prime modulus resembles multiplying two real numbers.

---

[1] Bad joke.

Now let's prove the lemma.

Proof Since $p|x$ if and only if $p|(-x)$, we may assume that $a, b \geq 0$.

Suppose, for a contradiction, that $ab \equiv 0 \pmod{p}$, $a \not\equiv 0 \pmod{p}$, and $b \not\equiv 0 \pmod{p}$.

Let $g = \gcd(b, p)$. As g divides p and p is prime, g must be either 1 or p. If it were p, since g divides b, we would have $p|b$, but we were told that $b \not\equiv 0 \pmod{p}$, so that cannot be the case. It follows that $g = 1$.

The Extended Euclidean Algorithm, applied to b, p, finds integers r, s such that $rb + sp = g = 1$. Therefore $rb - 1 = (-s)p$ and $rb \equiv 1 \pmod{p}$.

Now $(ab)r \equiv a(rb) \equiv a \cdot 1 \pmod{p}$. But $a \not\equiv 0 \pmod{p}$. So $(ab)r \not\equiv 0 \pmod{p}$. However, since $ab \equiv 0 \pmod{p}$, $(ab)r \equiv 0 \pmod{p}$, contradicting the fact that $(ab)r \not\equiv 0 \pmod{p}$.   ∎

A *corollary* is a theorem which is itself a simple consequence of another theorem. I know when I first saw "corollary" I couldn't understand why some theorems were called "theorems" and some were called "corollaries." How do you know which word to use? The answer is that you don't—the distinction is not formal. You could call all corollaries "theorems" and doing so would be correct. We use "corollary" instead just to emphasize the fact that the corollary follows easily and quickly from another theorem.

Corollary 2.1 *If p is prime, k is a positive integer, $m_1, m_2, \ldots, m_k$ are integers, and $p|(m_1 m_2 \cdots m_k)$, then there is some index i with $1 \leq i \leq k$ such that $p|m_i$.*

Proof If $k = 1$, the result is obvious. If $k = 2$, then Lemma 2.3 gives us the result when $a = m_1$ and $b = m_2$. So suppose $k \geq 3$.

Let $a = m_1$ and $b = m_2 m_3 \cdots m_k$; we have $p|(ab)$. Lemma 2.3 shows that p divides either a or b. In the former case, we are done. If not, then $p|b$.

Now let $a = m_2$ and $b = m_3 m_4 \cdots m_k$; $p|(ab)$. Either $p|a$ or $p|b$. In the former case, we are done.

In the latter case, repeat the argument. Eventually we run out of m_i's, so the process must terminate.   ∎

Here's a simple corollary of the corollary.

Corollary 2.2 *If p is a prime and $q_1, q_2, \ldots, q_k$ are all primes, and p divides $q_1 q_2 \cdots q_k$, then $p = q_i$ for some i.*

Proof Set $m_i = q_i$ and look at Corollary 2.1. The prime p must divide q_i for some i, but the only way one prime can divide another is for the two primes to be equal.∎

By the way, the really correct way to do this argument is through induction. We're finally ready to prove the uniqueness of prime factorization.

Theorem 2.3 *Uniqueness of Prime Factorization. Suppose that n is a positive integer, r, s are nonnegative integers, $p_1, p_2, \ldots, p_r$ are primes with $p_1 \leq p_2 \leq \cdots \leq p_r$ and $n = p_1 p_2 \cdots p_r$, $q_1, q_2, \ldots, q_s$ are primes with $q_1 \leq q_2 \leq \cdots \leq q_s$ and $n = q_1 q_2 \cdots q_s$. Then the two factorizations are the same. That is, $r = s$ and $p_1 = q_1, p_2 = q_2, \ldots, p_r = q_r$.*

Proof We do a proof by contradiction. If the statement is not true for every positive n, then it is false for at least one such n. Let the *smallest* value of n which has two truly different prime factorizations be called n also. (I hope this is not confusing.)

Clearly 1 has only one prime factorization, so $n > 1$. Let the factorizations of n be $n = p_1 p_2 \cdots p_r$ (with $p_1 \leq \cdots \leq p_r$) and $n = q_1 q_2 \cdots q_s$ (with $q_1 \leq \cdots \leq q_s$). We may assume $r \leq s$, since we can interchange the p's and q's if necessary to make it true.

Since the factorizations are distinct, there are two cases:

1. $p_1 = q_1, p_2 = q_2, \ldots, p_r = q_r$. We must have $s > r$ for otherwise the factorizations would be identical. But then $q_1 q_2 \cdots q_s$, which is supposed to be n, is larger than $p_1 p_2 \cdots p_r$, which equals n. So this case cannot happen.
2. There is some q_i, $1 \leq i \leq r$, which differs from p_i. Let j be the smallest such i. There are two cases:
   (a) $j \geq 2$. Then $p_1 = q_1$. Let $n' = n/p_1 = n/q_1$. Then we have two *different* prime factorizations for n', specifically, $n' = p_2 \cdots p_r = q_2 \cdots q_s$, contradicting the fact that n was the *smallest* positive integer having two distinct prime factorizations.
   (b) $j = 1$. (This is really the only interesting case.) Hence $p_1 \neq q_1$. One of these two is smaller than the other one. Let's say $p_1 < q_1$; the other case is similar. So $p_1 | n$ because $n = p_1 p_2 \cdots p_r$. But also $n = q_1 q_2 \cdots q_s$ so $p_1 | (q_1 q_2 \cdots q_s)$. Now we get to use Corollary 2.1, which we proved precisely to permit us to use it right now. By Corollary 2.1, p_1 must divide some q_i. However, p_1 is less than all the q_i's (since $p_1 < q_1 \leq q_2 \leq q_3 \leq \cdots \leq q_s$), and no prime divides any larger prime, so this is impossible.   ■

We have (finally) shown that prime factorizations are unique.

2.3   Puzzle

Assume the earth is a perfect sphere whose equator has length 40,000 km (about 25,000 miles). Imagine wrapping a belt around the equator so that it's taut. Now add one meter (about three feet) to the belt's length and rearrange it so that it floats uniformly the same height above the equator everywhere. Would you be able to place a bowling ball between the earth and the belt? A baseball? A penny? A playing card?

Here are dimensions to help you with your calculation:

Bowling ball: diameter about 21.5 cm or 8.5 inches.
Baseball: diameter about 7.5 cm or 3 inches.
Penny: thickness 1.52 mm or 0.06 inches.
Playing card: thickness 0.2 mm or 0.008 inches.

2.4 Exercises

Exercise 2.1 Prove, for every positive integer m, that $\equiv$ (with modulus m) is reflexive, symmetric, and transitive. Thus, it is an equivalence relation.

Exercise 2.2 A *partition* of a set U is a collection of non-empty sets whose union is U and which are pairwise disjoint (i.e., the intersection of any two distinct sets is empty). Each of the sets is called a *part* of the partition:

(a) Show that given any equivalence relation $\sim$ on a set S, there is a partition of S with the property that for any two elements x, y, $x \sim y$ if and only x and y are in the same part of the partition.
(b) Consider the equivalence relation $x \equiv y \pmod 3$ for integers x, y. What is the partition of the set of integers associated with this relation?
(c) Consider any partition P of a set S. For any x, y in S, define $x \sim y$ if and only if x and y are in the same part of the partition P. Show that $\sim$ is an equivalence relation.

Exercise 2.3 Show how to find the gcd of a and b given prime factorizations of a and b.

Exercise 2.4

(a) Prove that the gcd of 154 and 663 is 1.
(b) Find integers r, s such that $r \cdot 154 + s \cdot 663 = 1$.

Exercise 2.5 * Let a, b be positive integers with greatest common divisor g. Prove that any common divisor d of a and b divides g. (It's not possible, e.g., that the common divisors of a and b are 1, 2, 3, 5, with $g = 5$.)

Exercise 2.6 Code up the Extended Euclidean Algorithm in Python. I recommend writing it recursively if you're familiar with recursive programs.

2.4.1 Hint

2.5 Use the Euclidean Algorithm.

Modular Arithmetic

3

3.1 Modular Arithmetic

Most of the numbers one sees in everyday life are rational numbers, which are just fractions. You may know about irrationals like π. All of these are real numbers. By now you know a lot about the real numbers. You know there are infinitely many real numbers; after all, there are infinitely many integers, and every integer is a real number. Later on, in Chap. 6, we will see that while there are infinitely many integers and infinitely many real numbers, in a very real sense (pun intended), there are *more* real numbers than integers.

In this chapter we will study a new kind of math, one in which there are only *finitely* many numbers. This sounds bizarre already. How can there be only finitely many numbers?

The funny thing is that you are already very familiar with this kind of math since you use it every day to tell time. On a clock, there are 12 hours, labeled 12, 1, 2, 3, 4, 5, 6, 7, 8, 9, 10, 11. Forget about minutes and seconds; just think about hours. It will be convenient to replace the 12 by 0, so let's do so and report hours on a scale of 0, 1, 2, ..., 11.

What time is it 7 hours after 10:00? Surely it is not 17:00. It is 5:00. Well that's refreshing. You've convinced yourself for years that $10 + 7 = 17$. It must seem new and refreshing, albeit possibly disturbing, to have $10 + 7 = 5$. It's nice to shake things up occasionally.

Let's continue with some more examples. What time is it 3 hours after 6:00? That one's easy: it's 9:00. How about 17 hours before 3:00? Well, 12 hours before 3:00 was also 3:00, and 5 hours before 3:00 was 10:00.

How about 39 hours after 9:00? No, not 48:00, but that's a good start. The time 48:00 is the same as 0:00 (which we call 12:00 in the USA in ordinary life), because times that differ by a multiple of 12 hours are the same, and 48:00 and 0:00 differ by 48 hours, which is a multiple of 12.

© The Author(s), under exclusive license to Springer Nature Switzerland AG 2023   25
H. Karloff, *Mathematical Thinking*, Compact Textbooks in Mathematics,
https://doi.org/10.1007/978-3-031-33203-6_3

Let's try doing the same thing for minutes now, rather than hours. Suppose it is 23 minutes past the hour now. Ignoring the hour, how many minutes past the hour will it be in 72 minutes? The number of minutes past the hour will always be in $\{0, 1, 2, 3, \ldots, 59\}$, precisely because there are 60 minutes in an hour. Seventy-two minutes after 23 minutes past some hour, it will be $23 + 72 = 95$ minutes past that hour, or 35 minutes past the next hour, so the answer is "35 minutes past the hour."

On a piano, the names of the white keys repeat cyclically in the order A, B, C, D, E, F, G. (Traditionally, one starts from C, but let's simplify things by starting from A.) Convert the sequence A, B, C, D, E, F, G to the sequence $0, 1, 2, 3, 4, 5, 6$. What white key is 33 white keys above a C? What white key is 33 keys below a C? What white key would be 4001 white keys above a C, if pianos were exceptionally long? How about 4001 below?

There are 7 days every week. In the USA, we say the week starts on Sunday. Some countries start the week on Monday or Saturday. Starting on Sunday, the days of the week are Sunday, Monday, Tuesday, Wednesday, Thursday, Friday, and Saturday. Let's call these 7 days, in order, days 0, 1, 2, 3, 4, 5, 6, Sunday being day 0. Here's a natural problem. If today is Tuesday, what day will it be in 10 days? What day was it 10 days ago? What day of the week will it be in 4001 days? What day of the week was it 4001 days ago?

As you see, whenever there is a repeating pattern, we get the same structure.

In all these examples, we discussed how to add or subtract numbers. We will also discuss *multiplying* two numbers. In the clock example, it's hard to see why one would want to multiply 3:00 by 5:00, but we will do so anyway.

This kind of arithmetic, which only uses the remainders after you divide by some number m, is called *modulo-m* or *mod-m* arithmetic.

Recall this definition from Chap. 2.

Definition 3.1 Where a, b, m are integers and m is positive, we say that $a \equiv b$ (mod m) if $a - b$ is a multiple of m.

Let's prove a simple theorem which shows that in doing addition, subtraction, and multiplication mod m, any two numbers differing by a multiple of m are essentially this same. For example, if you're doing addition, subtraction, or multiplication with a modulus of $m = 10$, then anywhere you see a 1, you can replace it with an 11, a 21, a 31, $\ldots$, or a -9, a -19, a -29, etc. They're all the same. For example, if $m = 10$, then $337 \cdot 462 \equiv 7 \cdot 2$ (mod 10), because $337 \equiv 7$ (mod 10) and $462 \equiv 2$ (mod 10). This is because for nonnegative integers x, the last digit of x tells us what x is equivalent to mod 10. We have $7 \cdot 2 = 14 \equiv 4$ (mod 10). So $337 \cdot 462 \equiv 4$ (mod 10).

Theorem 3.1 *Suppose that m is a positive integer and that a, a', b, b' are integers such that $a' \equiv a$ (mod m) and $b' \equiv b$ (mod m). Then:*

1. $a' + b' \equiv a + b \pmod{m}$.
2. $a' - b' \equiv a - b \pmod{m}$.
3. $a'b' \equiv ab \pmod{m}$. *Here $a'b'$ and ab refer to multiplication.*

Proof Because $a' \equiv a \pmod{m}$, there is an integer r such that $a' - a = rm$. Similarly, there is an integer s such that $b' - b = sm$. Now:

1. $(a' + b') - (a + b) = (a' - a) + (b' - b) = rm + sm = (r + s)m$. Hence $(a' + b') - (a + b)$ is a multiple of m and hence $a' + b' \equiv (a + b) \pmod{m}$.
2. Similarly, $(a' - b') - (a - b) = (a' - a) - (b' - b) = rm - sm = (r - s)m$. Hence $(a' - b') - (a - b)$ is a multiple of m and hence $a' - b' \equiv (a - b) \pmod{m}$.
3. Last, $a'b' - ab = (a + rm)(b + sm) - ab = ab + rmb + asm + rmsm - ab = m(rb + as + rsm)$. Since $rb + as + rsm$ is an integer, $a'b' \equiv ab \pmod{m}$, and we are done. ∎

This theorem tells us that when doing addition, subtraction, and multiplication mod m, it's *only* the remainder when dividing by m that counts. As far as addition, subtraction, and multiplication go, if $m = 10$, then $\ldots, -29, -19, -9, 1, 11, 21, \ldots$ are all the same. We call them *equivalent mod 10*. They're all equivalent mod 10, so why don't we take a single one of them as a representative to represent them all? A good candidate for "representative" is the smallest nonnegative one. In the example mentioning $\ldots, -29, -19, -9, 1, 11, 21, \ldots$, we'll just pick 1 as the representative. Similarly, $\ldots, -28, -18, -8, 2, 12, 22, \ldots$ are all equivalent mod 10, and from them we pick 2 as the representative. Among $\ldots, -27, -17, -7, 3, 13, 23, \ldots$, we pick 3 as the representative. You get the idea. The representative is just the remainder when any one of the numbers is divided by m, that is, it's the number mod m.

By the way, all of this refers only to addition, subtraction, and multiplication, not necessarily to other operations such as exponentiation. For example, you can check that $2^{10} = 2 \cdot 2 \cdot 2 \cdot 2 \cdot 2 \cdot 2 \cdot 2 \cdot 2 \cdot 2 \cdot 2 = 1024$. Take $m = 10$ so that $10 \equiv 0 \pmod{m}$. However, $2^{10} = 1024 \equiv 4 \pmod{m}$, while $2^0 = 1 \equiv 1 \pmod{m}$, and 4 is not 1. You cannot replace the 10 in the exponent by a 0 and expect to get the same answer mod m.

How about division? The astute reader has probably been asking him- or herself, "What happened to division?" We've only discussed addition, subtraction, and multiplication. We'll discuss division in detail later. The situation is particularly interesting when m is a prime. But remember that you can't always divide two integers, for example, 1 by 2, and get an integer. More formally, there is no integer x such that $2 \cdot x = 1$. If you divide a pizza in half, each person doesn't get an integral quantity of pizza. That's why rational numbers were invented, though you have to wonder if the ancient person who invented rational numbers liked pizza. Maybe the pizza back then was so bad that he or she tried to give away half of it, and had to invent rational numbers in order to do so.

The same thing is happening here (not about the bad pizza; just about the inability to divide 1 by 2). Take $m = 4$. There are four numbers (representatives) when one does arithmetic mod 4: 0, 1, 2, 3. Is there an integer x such that $(2 \cdot x) \bmod 4 = 1$? To find out, let's try all the four x's. (We don't have to check any other x's. We don't have to check 4, 5, 6, or 7, for example.)

- $x = 0$: $(2 \cdot x) \bmod 4$ is $(2 \cdot 0) \bmod 4 = 0$, and 0 is not 1.
- $x = 1$: $(2 \cdot x) \bmod 4$ is $(2 \cdot 1) \bmod 4 = 2$, and 2 is not 1.
- $x = 2$: $(2 \cdot x) \bmod 4$ is $(2 \cdot 2) \bmod 4 = 0$, and 0 is not 1.
- $x = 3$: $(2 \cdot x) \bmod 4$ is $(2 \cdot 3) \bmod 4 = 2$, and 2 is not 1.

Thus there is no integer x such that $(2 \cdot x) \bmod 4 = 1$.

So far, if we compare the integers mod m to the rationals, we see a lot of similarities. For example, you can add two integers mod m to get another integer mod m. Similarly, you can add two rationals and get another rational. The same correspondence happens with multiplication: just as you can multiply any two integers mod m and get another integer mod m, you can multiply any two rationals and get another rational.

The set of rational numbers together with the operations of addition, subtraction, multiplication, and division is called a *field*, if those operations preserve certain properties. I won't bother you with the details of these properties, honestly because they're pretty boring. You can find the details if you search for "mathematical field" on the web. I remember when I first heard of the term "field" I couldn't understand where the name came from. Was it a reference to a baseball field? A grassy field? A soccer field? I was confused as to the origin of the name "field" back when I learned the term, and I am just as confused today as I was then. To me the name made no sense then and it makes no sense today. But the names are meaningless anyway.

The integers mod m *kind of* resemble the rationals. You can add, subtract, multiply, and—whoops, I forgot "divide." That's the one thing missing, the one thing that prevents arithmetic mod m from being a field. I can always divide any rational by any (nonzero) rational. However, as we saw a moment ago, in mod 4 arithmetic, you cannot always divide, since you cannot divide 1 by 2. Except for this "flaw," arithmetic mod m would be a field, just like the rationals.

Here's the key point. If you're doing modular arithmetic mod a *prime*, then you *can* always divide! Why? We'll get to that. But why am I excited about this? Because this is cool. It's a field, resembling the rationals, but it has only a *finite* number of elements. It's called a *finite field*. I doubt you've ever seen one before.

A *multiplicative inverse of x* is a number a such that $a \cdot x = 1$. We know that in the integers mod 4, there is no multiplicative inverse of 2. This phenomenon cannot happen if the modulus is prime.

Theorem 3.2 *Let m be a positive integer and let $1 \leq x \leq m - 1$. Then if $\gcd(x, m) = 1$, then x has a multiplicative inverse mod m. In particular, if m is prime, then since $\gcd(x, m) = 1$, x has a multiplicative inverse.*

If $\gcd(x, m) > 1$, then x does not have a multiplicative inverse mod m.

The proof follows easily from the Extended Euclidean Algorithm.

Proof Let $g = \gcd(x, m)$. Remember from the Extended Euclidean Algorithm that there are integers r, s such that $rx + sm = g$. Suppose that $g = 1$. Then there are integers r, s such that $rx + sm = 1$, and hence $rx - 1 = (-s)m$. But this means that $rx \equiv 1 \pmod{m}$. Therefore –and this is basically the same thing– $(r \bmod m)x \equiv 1 \pmod{m}$. Therefore $r \bmod m$ is the multiplicative inverse of x when doing mod m arithmetic.

Now suppose that $g > 1$. By the definition of gcd, $g|x$ and $g|m$. Suppose that a multiplicative inverse of x exists; let's call it r. Then $rx \equiv 1 \pmod{m}$. Therefore m divides $rx - 1$. That means there is an integer k such that $km = rx - 1$. However, $g|m$, so $g|(km)$. Since $km = rx - 1$, $g|(rx - 1)$. Since $g|x$, $g|(rx)$. But $g > 1$, so g can't divide both rx and $rx - 1$. That is impossible, so no multiplicative inverse of x can exist. ∎

Let's see some examples. Take $m = 7$, which of course is prime:

1. The multiplicative inverse of 1 is 1, because $(1 \cdot 1) \bmod 7 = 1$. (Boring.)
2. The multiplicative inverse of 2 is 4, because $(4 \cdot 2) \bmod 7 = 1$.
3. The multiplicative inverse of 3 is 5, because $(5 \cdot 3) \bmod 7 = 1$.
4. The multiplicative inverse of 4 is 2, because $(2 \cdot 4) \bmod 7 = 1$.
5. The multiplicative inverse of 5 is 3, because $(3 \cdot 5) \bmod 7 = 1$.
6. Last, the multiplicative inverse of 6 is 6, because $(6 \cdot 6) \bmod 7 = 36 \bmod 7 = 1$.

Here I've magically pulled the inverses out of a hat, but remember, the Extended Euclidean Algorithm shows you how to find inverses. You can even find multiplicative inverses when the modulus m has 1000 digits!

I won't go into details, but the fact that a multiplicative inverse of x exists if, and only if, $\gcd(x, m) = 1$, is used in the RSA algorithm for cryptography, which I mentioned in Chaps. 1 and 2. It is a very basic and important fact.

You may be asking yourself, "He showed us how to find the multiplicative inverse of x mod m. That's like the rational number $1/x$ or, say, $1/4$ if $x = 4$. But what if I have a larger numerator? What is the modular-arithmetic analogue of $3/4$?"

That's easy. In the rationals, to get $3/4$, you multiply $1/4$ by 3. You do the same thing with modular arithmetic. For example, if $m = 7$, we know that the inverse of 4 is 2. Then the analogue of $3/4$ is $(3 \cdot 2) \bmod 7 = 6$. Just as $4 \cdot (3/4) = 3$ in the rationals, mod 7 we have $4 \cdot (3/4) = (4 \cdot 6) \bmod 7 = 24 \bmod 7 = 3$, as it should be.

Now I'm going to present another fascinating fact about modular arithmetic, again when the modulus is prime. Again, this fact is useful in cryptography. Before we discuss it, here's an aside. Many years ago,[1] the branch of math called *number theory* that deals with the integers was considered to be beautiful and fun to do, but purely impractical. Why would anyone need number theory in the real world? Who

---

[1] when I studied math as an undergrad.

would care so much about the beautiful results of number theory? Who would care that the set of integers mod a prime defines a finite field? Were a lot of people wrong. Modern cryptography, in a sense, *is* number theory. To do cryptography these days, it really helps to know some number theory. The lesson is that you never know what will turn out to be useful in a few years.

Now let's turn to another fascinating result regarding mod-m arithmetic when m is a prime. Take $m = 7$ again. Now start with 1, and keep multiplying by 3 repeatedly:

1. 1
2. $(1 \cdot 3) \bmod 7 = 3.$
3. $(3 \cdot 3) \bmod 7 = 9 \bmod 7 = 2.$
4. $(2 \cdot 3) \bmod 7 = 6 \bmod 7 = 6.$
5. $(6 \cdot 3) \bmod 7 = 18 \bmod 7 = 4.$
6. $(4 \cdot 3) \bmod 7 = 12 \bmod 7 = 5.$
7. $(5 \cdot 3) \bmod 7 = 15 \bmod 7 = 1.$
8. $(1 \cdot 3) \bmod 7 = 3 \bmod 7 = 3.$

You can see now that the sequence will just cycle through the numbers 1, 3, 2, 6, 4, 5 forever. Notice that, except for 0, these are *all* the integers mod 7.

Let's see what happens if we start with 2 instead of 3:

1. 1
2. $(1 \cdot 2) \bmod 7 = 2.$
3. $(2 \cdot 2) \bmod 7 = 4 \bmod 7 = 4.$
4. $(4 \cdot 2) \bmod 7 = 8 \bmod 7 = 1.$

You can see that you're about to start repeating, though, this time, you haven't hit all the nonzero integers mod 7. You've missed 3, 5, 6.

Definition 3.2 Let p be a prime. Say the set of nonzero integers mod p is *cyclic* if there is a g between 1 and $p-1$ such that, if you start with 1 and multiply repeatedly by g, you get *all* the nonzero integers mod p. Such a g is called a *generator* of the set of nonzero integers mod p.

(We're using the letter "g" here to stand for "generator," not "gcd" as we used "g" to stand for before.)

We already saw that the set of nonzero integers mod 7 is cyclic. Let's try $p = 11$ instead. Which number g should we try? Clearly 1 won't work. Why don't we try $g = 2$? You, the reader, can verify that, starting from 1 and repeatedly multiplying by 2 mod 11, one gets this sequence, 1, 2, 4, 8, 5, 10, 9, 7, 3, 6, which contains all the integers between 1 and 10.

Since the set of nonzero integers mod 7 and the set of nonzero integers mod 11 were both cyclic, you can probably guess where I'm going with this.

Theorem 3.3 *For any prime p, the set of nonzero integers mod p is cyclic. That is, there is some g, $1 \leq g \leq p - 1$, such that $\{g^0 \bmod p, g^1 \bmod p, g^2 \bmod p, \ldots, g^{p-2} \bmod p\} = \{1, 2, 3, 4, \ldots, p - 1\}$.*

I apologize for this is one of those unhappy moments in a math-book author's life when he or she cannot give a proof. Multiple proofs are known, but none is simple enough for me to include here. I'm sorry, but this is one of those unfortunate times when you'll have to trust me.

By the way, the theorem says that *some* g between 1 and $p - 1$ must work. Trying even *one* g naively by computing its powers mod p would take billions of years, if p had 1000 digits (so $p > 10^{999}$), and, in this case in addition, there are far too many possible g's to try them all. For cryptography, better methods are known for the somewhat different version of the problem that is needed.

There is one salient point shared in common by all the proofs: they don't give a *fast* algorithm for finding the generator. What does this mean? The theorems do prove that, given a prime p, there is always some g between 1 and $p - 1$ such that the powers of g mod p give $1, 2, 3, \ldots, p - 1$, in some order. This means that the naive method of just trying potential generators g between 1 and $p - 1$ and seeing if powers of g generate all integers between 1 and $p - 1$ is guaranteed to work. This is a fine method for small primes, like those with, say, six digits, but it fails miserably for the really large primes, like those with hundreds of digits, which are used in cryptography. This naive method is way too slow, both because there are too many g's to try and because to test even *one* g, one has to examine the sequence $1, g, g^2 \bmod p, g^3 \bmod p, g^4 \bmod p, \ldots, g^{p-2} \bmod p$, where $p - 2$ may have 1000 digits and hence would require more than 10^{999} multiplications.

3.2 Discrete Logarithms

Given a generator g mod p, we can compute $g^l \bmod p$. Starting with $l = 0$, the first $p - 1$ values (those for $l = 0, l = 1, l = 2, \ldots, l = p - 2$) will be exactly the set $\{1, 2, 3, \ldots, p - 1\}$. This means that if I give you an h in $\{1, 2, 3, \ldots, p - 1\}$, there is exactly one value of l in $\{0, 1, 2, \ldots, p - 2\}$ such that $g^l \bmod p = h$.

Definition 3.3 Given a prime p and a generator g of the nonzero elements mod p, the function which takes any element h of $\{1, 2, 3, \ldots, p - 1\}$ and returns the unique exponent l in $\{0, 1, 2, \ldots, p - 2\}$ with $g^l \bmod p = h$ is called the *discrete logarithm* or *discrete log* function mod p, relative to generator g.

The "log" in the name comes from the fact that you're given the value of an exponential function and you're returning the exponent, just as you do with normal logs. The "discrete" part comes from the fact that there are only finitely many nonzeros mod p. Personally, I find complicated names like "discrete log" intimidating. Just remind yourself that *names don't matter in math*.

You definitely need some examples.

Examples. We saw above that for $p = 7$, $g = 3$ is a generator. Since $3^3 \bmod 7 = 6$, the discrete log of 6 mod 7 relative to generator 3 is the exponent, 3. We saw also that $g = 2$ is a generator for $p = 11$, with the powers of 2 being, in order, $1, 2, 4, 8, 5, 10, 9, 7, 3, 6$. Hence the discrete log of 3 mod 11 relative to generator 2 is 8, and the discrete log of 9 mod 11 relative to generator 2 is 6.

Computing discrete logs appears to be a difficult problem. No one has published any fast algorithm that, given a generator g of the set of nonzero integers mod p, will take a nonzero x and produce the discrete log of x mod p relative to generator g. Here, when I say "fast," I mean one that can feasibly work on 1000-digit integers. By the way, "appears" in "computing discrete logs appears to be a difficult problem" has two senses: first, conceivably someone knows a fast algorithm but hasn't revealed it (this is unlikely), and second, there probably doesn't *exist* a fast algorithm, but no one's been able to prove that yet.

Because computing discrete logs appears to be hard, a cryptosystem (a method for encrypting and decrypting messages) relying on the difficulty of computing discrete logs exists. If you could compute discrete logs mod 1000-digit primes, you could crack the cryptosystem, read encrypted e-mails, and maybe even steal lots of money, not that I would encourage you to do so.

3.3   Fermat's Little Theorem

Here's a fun little theorem, known as Fermat's Little Theorem. (Being French, Fermat would have pronounced his name "Fair-ma.")

Theorem 3.4 *Fermat's Little Theorem. If p is prime and a is not a multiple of p, then $a^{p-1} \equiv 1 \pmod{p}$.*

For example, since 4001 is prime, $10^{4000} \equiv 1 \pmod{4001}$.

Proof It is an important fact that if you multiply all $p - 1$ elements of the set $\{1, 2, 3, \ldots, p - 1\}$ by a and reduce mod p, you get the same set back (though probably in a different order). For example, if $p = 11$ and $a = 2$, we get $\{2, 4, 6, 8, 10, 12 \equiv 1, 14 \equiv 3, 16 \equiv 5, 18 \equiv 7, 20 \equiv 9\}$. Here's the proof. Clearly multiplying $1, 2, 3, \ldots, p - 1$ by a and reducing mod p gives us $p - 1$ numbers $a, 2a, 3a, \ldots, (p - 1)a$ between 1 and $p - 1$ inclusive. But are they all different? If they're all different, then they must be $1, 2, 3, \ldots, p - 1$ in some order. So suppose, for a contradiction, $ax \equiv ay \pmod{p}$ even though $x \not\equiv y \pmod{p}$. Then $a(x - y) \equiv 0 \pmod{p}$ even though $a \not\equiv 0$ and $x - y \not\equiv 0$. But since p is prime, we know that if p divides a product of two integers, it must divide one or the other, so this is a contradiction. This means that $\{a \bmod p, (2a) \bmod p, (3a) \bmod p, \ldots, ((p - 1)a) \bmod p\} = \{1, 2, 3, \ldots, p - 1\}$.

Now, let's continue. Let $M = [1 \cdot 2 \cdot 3 \cdots (p - 2) \cdot (p - 1)] \bmod p$. Since $\{a \bmod p, (2a) \bmod p, (3a) \bmod p, \ldots, ((p - 1)a) \bmod p\} = \{1, 2, 3, \ldots, p -$

1}, $(a \bmod p)((2a) \bmod p)((3a) \bmod p) \cdots ((p - 1)a \bmod p) \equiv M$. But this is $a^{p-1}M$. Hence $M \equiv a^{p-1}M$. However, since M is not 0 mod p, it has a multiplicative inverse mod p. Hence we can "divide" by M, getting

$$1 \equiv a^{p-1} \pmod{p},$$

and we are done.   ∎

3.4   The Chinese Remainder Theorem

Definition 3.4 Two nonnegative integers, not both 0, are *relatively prime* if their greatest common divisor is 1.

Here's a theorem, known to the Chinese in the first century CE, which I've always liked.

Theorem 3.5 *The Chinese Remainder Theorem. Let m_1, m_2 be relatively prime positive integers. Let $0 \le a_1 < m_1$ and $0 \le a_2 < m_2$. Then the system $x \equiv a_1$ (mod m_1), $x \equiv a_2$ (mod m_2) has a solution x in $\{0, 1, 2, \ldots, m_1 m_2 - 1\}$.*

Furthermore, the proof is constructive in that it suggests an algorithm to find the desired x.

Let's take an example before we prove the theorem, so I can show you the beauty of the theorem. Take $m_1 = 3$, $m_2 = 5$. Look at all the integers x in $\{0, 1, 2, \ldots, 3 \cdot 5 - 1\} = \{0, 1, 2, \ldots, 14\}$:

x	x mod 3	x mod 5
0	0	0
1	1	1
2	2	2
3	0	3
4	1	4
5	2	0
6	0	1
7	1	2
8	2	3
9	0	4
10	1	0
11	2	1
12	0	2
13	1	3
14	2	4

The key point (and this is not implied by the short form of the Chinese Remainder Theorem I've given you) is that each possible pair of numbers in columns 2 and 3 *appears exactly once.* This is not a fluke. It always happens, provided only that m_1 and m_2 are relatively prime.

Now it's time for the proof.

Proof The congruence $x \equiv a_1 \pmod{m_1}$ is equivalent to this equation: $x = a_1 + m_1 y$ for some integer y. Hence the second congruence is the same as $a_1 + m_1 y \equiv a_2 \pmod{m_2}$. Subtract a_1 from both sides. We need to solve $m_1 y \equiv (a_2 - a_1) \pmod{m_2}$. Now we use the fact that m_1 and m_2 are relatively prime to infer that there is an inverse m_1' of m_1 mod m_2 (and Theorem 3.2 tells us how to find the inverse). Multiply both sides by m_1'; we get $y \equiv m_1'(a_2 - a_1) \pmod{m_2}$. It follows that $y = m_1'(a_2 - a_1) + m_2 z$ for some integer z. Hence:

$$x = a_1 + m_1 y = a_1 + m_1(m_1'(a_2 - a_1) + m_2 z) = a_1 + m_1 m_1'(a_2 - a_1) + m_1 m_2 z.$$

It is easy to check, using the fact that $m_1 m_1' \equiv 1 \pmod{m_2}$, that if we set $x = a_1 + m_1 m_1'(a_2 - a_1)$, then $x \equiv a_1 \pmod{m_1}$ and $x \equiv a_2 \pmod{m_2}$. (If the solution we found is not in $\{0, 1, 2, \ldots, m_1 m_2 - 1\}$, add $z(m_1 m_2)$ for some integer z to put it in the set; doing so doesn't change the fact that it is a solution.) ∎

In fact, here's an extension which we won't prove.

Theorem 3.6 *Suppose $m_1, m_2, \ldots, m_n$ are positive integers such that for all $1 \le i < j \le n$, m_i and m_j are relatively prime. Let $a_1, a_2, \ldots, a_n$ be any integers with $0 \le a_i < m_i$ for all i.*

Let $M = m_1 m_2 \cdots m_n$. Then the system of congruences

$$x \equiv a_i \pmod{m_i}$$

for all i has a solution. In fact, there is exactly one solution x^* in the set $\{0, 1, 2, \ldots, M - 1\}$. The set of all solutions x in the integers is the set of all integers of the form $x^* + Mz$ for integers z.

3.5 Puzzle

On one section of track sit two locomotives facing each other, 100 kilometers apart. At noon, both trains start moving toward each other. The first train T_1 travels from the left at 20 kilometers per hour; the second train, T_2, travels from the right at 30 kilometers per hour. At noon, a fly departs from the front of T_1 and, flying rightward at 40 kilometers per hour, flies until it reaches the front of T_2. It then turns around instantaneously and, still flying at 40 km/h but leftward, flies toward T_1. Once it reaches T_1, it turns around instantaneously and flies rightward at 40 km/s until it reaches T_2, at which time it turns around.

Eventually, after infinitely many reversals, the fly is crushed between the two trains when they collide. How far did the fly fly before being crushed?

3.6   Exercises

Exercise 3.1 * Let m be 2 or 5. Prove via induction on n that a nonnegative integer n is a multiple of m if and only if its last digit is a multiple of m. That is, show that n is a multiple of 2 (i.e., is even) if and only if its last digit is 0, 2, 4, 6, or 8, and that n is a multiple of 5 if and only if its last digit is 0 or 5.

Exercise 3.2 * Let m be 3 or 9. Prove by induction on $n \geq 0$ that n is a multiple of m if and only if m is a divisor of the sum of the digits of n. For example, 38392839 is a multiple of 9 because $3 + 8 + 3 + 9 + 2 + 8 + 3 + 9 = 45$ is a multiple of 9.

Exercise 3.3 For any positive integer n, define $\phi(n)$ (read "fee of n") to be the number of positive integers less than n which are relatively prime to n. For example, $\phi(6) = 2$ because 1 and 5 are relatively prime to 6 but 2, 3, and 4 are not. $\phi(9) = 6$ because 1, 2, 4, 5, 7, 8 are relatively prime to 9 but 3, 6 are not:

(a) What is $\phi(n)$ if n is prime?
(b) What is $\phi(n)$ if $n = p_1 p_2$ where p_1 and p_2 are two *different* primes? Suppose that p_1 and p_2 are approximately $\sqrt{n}$. Is $\phi(n)$ a large fraction of n or a small fraction of n?
(c) What is $\phi(n)$ if $n = p^2$ where p is a prime?

Exercise 3.4 What is 3^{1000} (mod 11)? Use Python to compute 3^{1000} (mod 11) (this is 3**1000 % 11 in Python).

Exercise 3.5 Eleven is prime. By Theorem 3.3, $\{1, 2, 3, \ldots, 10\}$ is cyclic under multiplication mod 11. Find all generators of $\{1, 2, 3, \ldots, 10\}$.

Exercise 3.6 Theorem 3.4, Fermat's Little Theorem, shows that if p is a prime and $a \not\equiv 0$ (mod p), then $a^{p-1} \equiv 1$ (mod p). Prove that if p is a prime greater than 2 and $a \not\equiv 0$ (mod p), then $a^{(p-1)/2} \equiv 1$ (mod p) or $a^{(p-1)/2} \equiv -1$ (mod p).

Exercise 3.7 Let p be an odd prime. Let a be an integer, $0 \leq a < p$. Say that "a is a perfect square mod p" if there is an integer x such that $x^2 \equiv a$ (mod p).
How many perfect squares mod p are there?

Exercise 3.8 * Let p be an odd prime and suppose that g is a generator of the nonzero integers mod p. Let k be a nonnegative integer such that $\gcd(k, p-1) = 1$. (It's $p - 1$ here, not p.) Show that g^k is a generator of the nonzero integers mod p.
For example, if $p = 7$, the powers of 3 mod 7 are 1, 3, 2, 6, 4, 5, so 3 is a generator mod 7. $k = 5$ is relatively prime to $p - 1 = 6$, so 3^5 mod 7, which is 5, should also be a generator. The powers of 5 mod 7 are 1, 5, 4, 6, 2, 3 so indeed 5 is a generator of the nonzero integers mod 7.

Exercise 3.9 Suppose $p_1, p_2, \ldots, p_k$ are distinct primes and that $x \equiv 0 \pmod{p_i}$ for all i. Show that $x \equiv 0 \pmod{p_1 p_2 \cdots p_k}$.

Exercise 3.10 In Theorem 3.6, the hypothesis is that each pair of moduli is relatively prime. Maybe this condition is unnecessarily strong. Consider replacing the condition that all pairs are relatively prime by the condition that there is no common divisor $d > 1$ of all the moduli. Prove that the "theorem" with the modified hypothesis is false.

Exercise 3.11 Without a calculator or computer, solve this system:

$$x \equiv 25 \pmod{37}$$

$$x \equiv 39 \pmod{96}$$

Exercise 3.12 Show that the congruence $ax \equiv b \pmod{m}$ has a solution if and only if the greatest common divisor of a and m divides b.

Exercise 3.13 * An important operation is *modular exponentiation*, that is, given positive integers a, b, m, compute a^b mod m. Here a, b, m may have hundreds or even thousands of digits.

(a) Explain why one should not do this by computing a^b and then reducing modulo m.
(b) The next attempt is to do:
   1. $result = 1$
   2. For $i = 1$ to b, do:
      (a) $result = (result * a) \pmod{m}$
   Explain why this won't work.
(c) How *should* one compute a^b mod m?

Exercise 3.14 Write a Python program that takes an odd prime p and finds a generator of $\{1, 2, 3, \ldots, p - 1\}$.

Exercise 3.15 Write a Python program to generate discrete logs. Your program will take an odd prime p and a generator g. Assume p really is prime and g really is a generator. Can your program run feasibly on six-digit numbers? Ten-digit numbers? I won't even ask about 1000-digit numbers.

3.6.1   Hints

3.1 Start by proving by induction on n that, where l is n's last digit, $n \equiv l$ mod m.
3.2 Start by proving by induction on n that $n \equiv s$ mod m where s is n's sum of digits.

3.8 Use the fact that k has an inverse mod $(p - 1)$.

3.13 (c) First think about the case in which b is a large power of 2, like 2^{100}, and then think about general b's.

Irrationals

4

4.1   Rationals and Irrationals

Years ago when you started doing arithmetic, you knew only about the nonnegative integers 0, 1, 2, 3, 4, ... Maybe you were surprised to learn about negative integers. Do negative numbers really exist? They certainly exist as bank balances: a bank balance of -$5 means that you owe the bank $5.

Negative numbers were introduced into mathematics approximately 2000 years ago by the Chinese. Probably negative numbers were introduced to allow the subtraction of any number from any other; otherwise, what would $3 - 8$ be? If there were no negative numbers, then would $(3 - 8) + 8$ still be 3, after a temporary detour into never-never land?

To you, I'm sure by now that negative numbers seem perfectly natural and that you would object at the possibility of a number line that started at 0 and only proceeded rightward. Perhaps after some initial unease upon seeing negative integers, you made peace with the negative integers.

While learning addition, subtraction, and multiplication on the integers, everything was fine. . . until you learned about division. At the beginning, you were probably given division problems with integral results, like 8/4. The next step was probably dividing two integers and getting an integral quotient and a remainder. Still everything is fine, since you stayed within the integers. Then you learned about fractions. Whoa—a new kind of number! But maybe even this didn't seem unreasonable. After all, 1/2 is just the amount of pizza given to each of two people who share a pizza equally. Eventually you internalized fractions. You may have thought that all numbers were fractions, if you included improper fractions like 4/2.

H. Karloff, *Mathematical Thinking*, Compact Textbooks in Mathematics, https://doi.org/10.1007/978-3-031-33203-6_4

Because fractions are *ratios* of two integers, numbers which have fractional representations (such as 1/2) as the ratios of two integers are called *rationals*. Their denominators must be nonzero, since division by zero is not defined.

If all the numbers you've ever seen are rational, then it may not have crossed your mind to think about any other kinds of numbers.

The reason you've seen so few, or no, irrationals is that if you start with rationals, doing arithmetic on them always generates rationals, so that you'll never get *any* irrationals this way. This is very easy to prove.

Theorem 4.1 *If x and y are rationals, then $x + y$, $x - y$, and $x \cdot y$ are all rationals, and if $y \neq 0$, then so is x/y.*

Proof Since x is rational, there are integers a, b, with $b \neq 0$, such that $x = a/b$. Just the same, there are integers $c, d, d \neq 0$, such that $y = c/d$.

Now $x + y = a/b + c/d = (ad + bc)/(bd)$. Notice that $bd \neq 0$ since $b, d \neq 0$ and the product of two nonzero numbers is nonzero. Because a, d, b, c are all integers, so are $ad + bc$ and bd, so $(ad + bc)/(bd)$ is the ratio of two integers with the second one nonzero, and is hence rational.

Similarly, $x - y = a/b - c/d = (ad - bc)/(bd)$. By the same sort of reasoning, we infer that $ad - bc$ and bd are integers with $bd \neq 0$, and we conclude that $x - y$ is rational.

Now let's look at xy and x/y. We have $xy = (ac)/(bd)$ with ac, bd both integers with $bd \neq 0$. Hence xy is rational. Furthermore, if $y \neq 0$, then since $y = c/d$ with $c \neq 0$, $x/y = (a/b)/(c/d) = (a/b)(d/c) = (ad)/(bc)$, and $bc \neq 0$ (since $b, c \neq 0$); hence $(ad)/(bc)$ makes sense. Because a, d, b, c are integers, so are ad and bc, so x/y is rational. We are done. ∎

Are there any numbers which are not rational? Yes! These numbers are called *irrationals*. In fact, in a real sense, as we will see in Chap. 6, there are many *more* irrationals than rationals. Even further, if one were to pick a "random" real number, it would almost certainly be irrational. Among all the real numbers, the rational numbers, the ones you've been seeing for years, are a negligible fraction of the total. If the real numbers were a haystack containing a needle, the rationals would be the needle in that haystack and the irrationals the hay. Truly the rationals are an extremely small fraction of the reals.

Have you ever seen any irrational numbers? Indeed you have, if you've ever seen π, which is pronounced as is "pie." This strange-looking number is the ratio of the circumference of a circle (its perimeter) to its diameter. This ratio is the same for all circles.

EXPERIMENT. Take a mug and a tape measure if you have one, or a piece of string and a ruler if you don't. Wrap the tape measure or string circularly around the mug and measure its circumference. Then measure the mug's outside diameter, meaning that you should include the thickness of the mug. Divide the big number by the small number. If the mug's cross section were a perfect circle, and if you measured perfectly, you'd get π.

I just tried this myself. I got 10.4 inches and 3.25 inches, which gives an estimate of $10.4/3.25 = 3.2$ for π. Better approximations –which are worth memorizing– are 3.14 and 22/7. Remember, though, that these are just approximations.

You may have asked yourself, when first introduced to π, why did we give this important number such a strange name, which is a letter from the Greek alphabet? (It's funny to note that in Greek, the language from which the letter π –which we pronounce "pie"– originated, that letter is pronounced exactly as we pronounce the letter "p.") In math, people often use letters from the Greek alphabet. The first three letters of the Greek alphabet are α, which is written in English as "alpha" and pronounced "alfa"; β, which is written in English as "beta" and pronounced "baytuh"; and γ, written in English as "gamma" and pronounced as "gamma." When you see unfamiliar letters in math, it's a good bet they came from Greek. By the way, our word "alphabet" came from the first two letters, "α =alpha" and "β=beta," in the Greek alphabet.

One thing that always confused me when I was a kid was the fact that certain Greek letters have prescribed meanings. You know that π is the ratio of the circumference of a circle to its diameter. In fact, π is just one of the letters in the Greek alphabet. It means whatever you want it to mean. Saying $\pi = 3.14159\ldots$ is like saying $p = 3.14159\ldots\ldots$. Would I never be able to use p again, for any other purpose, if I did so? What a waste of a letter. π is simply a Greek letter. We can assign any number to it. For some reason, we've decided that the particular Greek letter π must equal $3.14159\ldots$ and that we cannot use it for anything else. Preposterous.

Let's return to the math. I know I was intimidated when I saw π as a kid. What is the strange symbol? Why didn't we just give this number a normal name? Well, we didn't have to use the Greek alphabet; we could have just called it, say, p. But why didn't we give it a normal, rational name, like 22/7 or $3.14 = 314/100$, for example? The answer is that we couldn't. There is *no* exact rational name for π, because π isn't rational! The circumference of a circle of radius one, π appears very naturally in mathematics—after all, we study perimeter of rectangles, so why shouldn't we study the circumference of a circle—but it defies our attempts to assign it the kind of name we were used to, because quite simply it isn't rational. How do people know that π is irrational? Proving a number x irrational seems like a remarkable feat in itself. There are infinitely many possible positive denominators q. For each one, the only numerator that can work is $p = q \cdot x$ (for if $p/q = x$, then $p = q \cdot x$). If for any positive integer q, $q \cdot x$ is an integer, then x is rational. In order for x to be irrational, $q \cdot x$ must not be integral, for every single one of the infinitely many possible q's. How is it possible to write a proof ruling out all of the infinitely many possible rational numbers without trying them all, one at a time? An impressive achievement indeed, it seems.

Unfortunately, proving that π is irrational is quite difficult, and you will not find the proof in this book. Instead, without too much effort, we will prove a couple other numbers to be irrational.

To get started, we have to review decimal representations of numbers. Probably you've learned that all fractions can be represented in decimal format. Every real number x as well has a decimal representation as a finite string of digits before a decimal point, followed by a finite or infinite string of digits after the decimal point. For example, 65.35 is the decimal representation of $65 + 35/100 = 6535/100$. This representation is *terminating* since the string after the decimal terminates. You know that other numbers, like $1/3$, have *nonterminating* representations, in that the string after the decimal point goes on forever: $1/3 = 0.3333333\ldots$. If a pattern repeats, as it does here, we often write a bar over the repeating pattern. Hence $1/3 = 0.33333333\ldots = 0.\overline{3}$. A decimal representation is called *repeating* if it consists of a finite sequence of digits containing a decimal point somewhere followed by an infinite number of copies of a nonempty finite block of digits, for example, $720.32\overline{45}$ (the "45" going on forever).

Since it gets tiresome to continue saying "finite or infinite," let's just tack on an infinite string of zeros to the end of each terminating decimal, so that 65.35 becomes instead $65.35\overline{0} = 65.3500000000000000000000000000000000000000\ldots$. That way, we can say that every terminating representation can be viewed as actually nonterminating. Let's insist that all of our decimal representations go on forever.

This suggests a subtle question. Can one number have two or more different representations? I don't mean one terminating and one nonterminating. I mean, can one number have two or more different nonterminating representations? The answer, unfortunately, is "yes," but barely. Can you find two different representations for the same number? Please, stop and think for a few minutes before proceeding.

We will need a simple theorem which you probably already know.

Theorem 4.2 *If* $-1 < r < 1$, *then the infinite sum* $1 + r + r^2 + r^3 + \cdots = 1/(1-r)$.

For example, if $r = 1/10$, then $1 + 0.1 + 0.1^2 + 0.1^3 + \cdots = 1/(1 - 1/10) = 1/(9/10) = 10/9$. Hence $0.1 + 0.1^2 + 0.1^3 + \cdots = 0.1111\ldots = 0.\overline{1} = -1 + 10/9 = 1/9$. If you have 9's instead of 1's, you get something 9 times as large: $0.999\ldots = 0.\overline{9} = 9 \cdot (1/9) = 1$.

A few words about this sum. Its value is *exactly* 1. It is not "very close" to 1. It is not "1 in the limit." It is *exactly* 1. It follows that the number 1 has two decimal representations: $1 = 1.\overline{0} = 0.\overline{9}$.

Similarly, $0.5 = 0.5\overline{0} = 0.500000000000000\ldots = 0.4\overline{9} = 0.499999999999999\ldots$ and $0.37 = 0.37\overline{0} = 0.3699999999999999\ldots$. However, that's the only kind of number with two or more different decimal representations.

Theorem 4.3 *Any number* x *with a decimal representation ending in infinitely many zeros (i.e., having a terminating representation) has exactly two representations, the other one ending in infinitely many nines. Every real number* x *which does not have a terminating representation has exactly one decimal representation.*

Proving this theorem is Exercise 4.11.

So what, you ask? Why are decimal representations useful in trying to prove a number irrational? Here's the key fact.

Theorem 4.4 *A real number is rational if, and only if, it has a repeating decimal representation.*

We will prove this theorem in the next section. It is not difficult.

Now, let's use this handy theorem to prove that irrational numbers exist (without relying on my word that π is irrational). In light of the theorem, all we have to do is find a real number which does not have a repeating decimal representation. How about this one, which we will call z:

$$z = 0.1101000100000001000000000000001\ldots.$$

Do you see the pattern? This number has a 1 in the lth digit after the decimal point if and only if $l = 1, 2, 4, 8, 16, 32, \ldots$, that is, if and only if l is a power of 2.

Theorem 4.5 *z is irrational.*

Proof Clearly z ends in neither infinitely many 0's nor infinitely many 9's, so it cannot have another decimal representation. It is not repeating. (If it repeated, it would either have zeros forever after some point, which it doesn't, or it would have evenly spaced repeating nonzero digits forever, which it doesn't.) Hence z is irrational. ∎

4.2   The Rationals Are the Reals with Repeating Decimal Representations

I still owe you a proof of Theorem 4.4.

Theorem 4.6 *Every fraction a/b for $b \neq 0$ has a repeating decimal representation.*

Proof We may assume that a and b are both nonnegative, for if either or both are nonnegative, we may strip off the sign(s) and introduce the correct sign on the quotient later. Now add a decimal point to the right of a and then add infinitely many zeros. Think about doing long division to divide a by b. After a while, we exhaust the digits of a and process the (very) long string of zeros following a. In each step of the long division algorithm, we divide b into some remainder, figure out the *single*-digit quotient, multiply b by the single-digit quotient, subtract the product from the remainder, *bring down a zero*, and continue. There are two important points here:

1. We always bring down a zero.
2. The remainder must be an *integer* between 0 and $b - 1$.

Since there are infinitely many zeros and only b remainders, eventually the same remainder will occur a second time. Once that happens, the algorithm will generate exactly the same one-digit quotients as before, forever, guaranteeing that the final decimal representation will repeat forever. We are finished. ∎

To see this proof in action, just divide 1 by 7 by hand.

So we have shown that any fraction has a repeating decimal representation. Is the *converse* true? The *converse* of a statement is the same statement "backward." It has nothing to do with sneakers.[1] In other words, the converse, which may or may not be true, even if the original statement is true, is, "Any number with a repeating decimal representation is a rational number." It's important to see that this is not the same as the original statement. The original statement is that every rational number has a repeating decimal representation. However, maybe some irrational numbers *also* have infinite repeating decimal representations! In fact, that cannot happen.

Define a *pure repeating decimal* to be one which starts with a 0 and decimal point and then has a block which repeats immediately and infinitely, for example, $0.\overline{120} = 0.120120120120\ldots$. It is easy to see that if every pure repeating decimal is rational, then so is every repeating decimal. For example, $3.12\overline{56} = 3 + \frac{12}{100} + \frac{1}{100} \cdot 0.\overline{56}$. If $0.\overline{56}$ is rational, then so is $3.12\overline{56}$.

The value of a pure repeating decimal is always easy to get. Here, I'll give you a few examples. See if you see the pattern:

$$0.4444444444\ldots = 4/9.$$

$$0.1313131313\ldots = 13/99.$$

$$0.234234234234\ldots = 234/999.$$

$$0.5607560756075607\ldots = 5607/9999.$$

See the pattern?

Theorem 4.7 *Any real number with a repeating decimal representation is rational.*

This is not hard to prove, but instead of proving the theorem, I will just illustrate it with an example, which will show you how to prove it. Consider the number, say, $12.\overline{56} = 12.56565656565656\ldots$. Let's call this number h. Then:

$$h = 12 + (0.56/1 + 0.56/100 + 0.56/100^2 + 0.56/100^3 + \cdots)$$
$$= 12 + .56(1 + 1/100 + 1/100^2 + 1/100^4 + \cdots).$$

Theorem 4.2 implies that the infinite sum

$$1 + r + r^2 + r^3 + \cdots$$

---

[1] Bad joke.

equals $1/(1-r)$, provided that $-1 < r < 1$. This means that $1 + 1/100 + 1/100^2 + \cdots = 1/(1 - 1/100) = 100/99$, because $r = 1/100$. Now that we've done the hard part, the rest is just mopping up:

$$h = 12 + 0.56 \cdot \frac{100}{99}$$

$$= 12 + \frac{56}{100} \cdot \frac{100}{99}$$

$$= 12 + \frac{56}{99}$$

$$= \frac{12 \cdot 99 + 56}{99},$$

which is clearly rational.

Notice in this example, the repeating pattern "56" had length 2, and the geometric series we needed, $1 + 1/100 + 1/100^2 + 1/100^3 + \cdots = 1 + 10^{-2} + (10^{-2})^2 + (10^{-2})^3 + \cdots$, had value $10^2/99 = 10^2/(10^2 - 1)$. It would be inspired, and even correct, to conjecture that for a repeating pattern of length l, the corresponding series

$$= 1 + \left(\frac{1}{10^l}\right) + \left(\frac{1}{10^l}\right)^2 + \left(\frac{1}{10^l}\right)^3 + \cdots$$

would have sum

$$\frac{1}{1 - \frac{1}{10^l}} = \frac{10^l}{10^l - 1},$$

which is surely rational. This means you can write down the sum "by inspection," which means "by glancing." For example, $0.\overline{347} = 0.347347347347\ldots = \frac{347}{999}$, because the sum is $\frac{347}{1000} \cdot \frac{1000}{999} = \frac{347}{999}$.

Even though I've not provided a formal proof, we've effectively finished the proof of Theorem 4.4 that the rationals are exactly those real numbers with a repeating decimal representation.

4.3 Square Roots

4.3.1 The Pythagorean Theorem

Have you seen right triangles? A *right triangle* is a triangle with a right angle; a *right angle* is an angle between two perpendicular line segments. The longest side of a right triangle comes with a complicated and difficult-to-pronounce name which comes to us from the ancient language of Latin: *hypotenuse*. (Long, complicated, foreign names always intimidated me. Why do we use such complicated names?

Because we think they make us look smart? I don't know, but we all do use complicated names.) The other two sides of a right triangle are called the *bases* or *legs*.

Now here's the question: if I give you the lengths of the legs of a right triangle, can you tell me how long the hypotenuse is? If the legs were of lengths 3 and 4, how long would the hypotenuse be?

EXPERIMENT. Take out some paper. Find a right angle and draw perpendicular legs of length 3 inches and 4 inches. Then take out your ruler and measure the hypotenuse. What did you get?

If you drew a perfect right triangle, whose legs were exactly 3 and 4 inches, then your hypotenuse would be exactly 5 inches long.

How do I know this? Because there's a very famous theorem, the *Pythagorean Theorem*, which tells you how to calculate the length of the hypotenuse from the lengths of the legs. It's called the Pythagorean Theorem –another one of those long, foreign-sounding math terms– because it is attributed to ancient Greek mathematician Pythagoras.

Theorem 4.8 *The Pythagorean Theorem. Suppose a right triangle has legs of lengths a and b and hypotenuse of length c. Then $c^2 = a^2 + b^2$. (In other words, $c \cdot c = a \cdot a + b \cdot b$.)*

For example, if $a = 3$ and $b = 4$, then the unknown c must satisfy $a^2 + b^2 = c^2$. But $a^2 = a \cdot a = 3 \cdot 3 = 9$ and $b^2 = 4 \cdot 4 = 16$, so $9 + 16 = c^2$. Since $9 + 16 = 25$, we know that c times itself is 25. The only positive number satisfying this equation is 5. (I inserted "positive" because there actually is another number, namely, -5, which works, since $(-5) \cdot (-5) = 25$ also, but since I've never seen a triangle with a negative edge length, we can rule out the -5.)

Another example is a right triangle with legs of lengths 5 and 12. Then we have $c^2 = 5^2 + 12^2 = 25 + 144 = 169$, so $c^2 = 169$. The only positive number c which satisfies $c^2 = 169$ is $c = 13$, so the hypotenuse has length 13.

EXPERIMENT. Draw yourself a right triangle with two legs of length 1. Measure the length of the hypotenuse, and write it down. We'll return to this later.

Now we will give a proof of the Pythagorean Theorem. Proofs of the Pythagorean Theorem must use some facts about the Euclidean plane on which the triangle is drawn. We will use these facts about shapes in the Euclidean plane:

1. The area of a square whose side length is c is c^2.
2. The area of a right triangle with legs a and b is $(1/2)ab$.
3. The sum of the angles in a triangle is $180°$.
4. The sum of the angles along a line is $180°$.

Look at Fig. 4.1. We started with a square of side length $a + b$ and marked a point at distance a from each corner, as in the picture. The remaining part of each of the square's sides has length b. Since the corner of a square is a right angle, we have

Fig. 4.1 Figure for the proof
of the Pythagorean Theorem

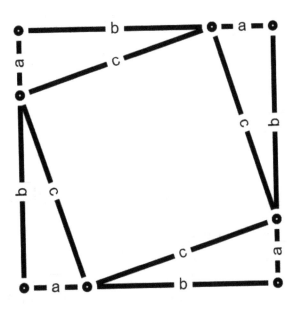

drawn four right triangles, each of area $(1/2)ab$, and therefore the total area of all four triangles is $4 \cdot (1/2)ab = 2ab$.

In each of the four right triangles, let the angle *opposite* the side of length a measure, say, A degrees, and let the angle opposite the side of length b measure, say, B degrees. Since the sum of the angles in a triangle is $180°$, and the right angle measures $90°$, we have $A + B + 90 = 180$, and so $A + B = 90$.

In the middle of Fig. 4.1, we see a quadrilateral that looks an awful lot like a square. In fact, it must be a square. Its side lengths are all c, but how do we know that the angles are all $90°$? Pick any one angle and say its measure, in degrees, is x. The three angles at the corner add up to $180°$: $A + x + B = 180$. But we showed that $A + B = 90$, which means that $x = 90$, and hence corresponds to a right angle. Hence the quadrilateral has four equal sides and four right angles so really *is* a square.

Now we play with areas. The area of the $c \times c$ square in the middle is $c \cdot c = c^2$. The total area of the four triangles is $2ab$.

What is the area of the large $(a+b) \times (a+b)$ square? It's $(a+b) \cdot (a+b)$. Now what is that? I specifically remember getting this wrong myself once when I was in ninth grade –I am not making this up– so don't be upset if you think it's a^2+b^2. It's not. It's $a^2+2ab+b^2$. Why? By the distributive law, for any number y, $(a+b) \cdot y = ay + by$. Now set $y = (a+b)$. We have $ay + by = a(a+b) + b(a+b)$, which, by two more applications of the distributive law, is $a^2 + ab + ba + b^2 = a^2 + 2ab + b^2$.

The area of the big square (which equals $a^2 + 2ab + b^2$) equals the area of the little square (c^2) plus the area of the four triangles ($2ab$ in total). So we have:

$$a^2 + 2ab + b^2 = c^2 + 2ab.$$

We subtract $2ab$ from both sides and conclude that $a^2 + b^2 = c^2$, as we were trying to prove.   ∎

4.3.2   Square Roots

In a square of side length 1, how long is a diagonal? To put it another way, in a right triangle with two legs of length 1, how long is the hypotenuse? We have $a = b = 1$, so if c is the length of the hypotenuse, then $1^2 + 1^2 = c^2$. Clearly $1^2 = 1$ and so $c^2 = 2$. *What is c?* You measured the hypotenuse of such a triangle recently. What did you get, about 1.4? Let's apply the Pythagorean Theorem. The two legs are of lengths $a = 1$ and $b = 1$, and $1^2 + 1^2 = 1 \cdot 1 + 1 \cdot 1 = 1 + 1 = 2$. By the Pythagorean Theorem, where c is the length of the hypotenuse, $c^2 = a^2 + b^2 = 2$; hence $c^2 = 2$. What is c? It's approximately 1.4, since $1.4^2 = 1.96$. A better approximation is 1.41, whose square is 1.9881. Even better, 1.414, whose square is 1.999396. You would do even better with 1.414213562373, whose square is 1.99999999999973116139. For lack of a better name, let's call the nonnegative number whose square is 2 the *square root of 2* and write it as $\sqrt{2}$. In general, if x is a real number, the *square root of x* is the nonnegative real number whose square is x. For example, the square root of 9 is 3. The square root of 5 is approximately equal to 2.236067977.

If x is negative, is there any number y such that $y^2 = x$? Well, no, since regardless of whether y is positive, negative, or 0, $y^2 \geq 0$, so cannot equal x. So there are no real square roots of negative numbers.

Given the title of this chapter, you're probably surmising that square roots are irrational. Now this isn't always true, obviously, since $\sqrt{0} = 0$, $\sqrt{1} = 1$, $\sqrt{4} = 2$, $\sqrt{9} = 3$, for example, are all integers and hence rational. But how about that pesky $\sqrt{2}$? There is a miraculously simple proof that $\sqrt{2}$ is irrational, given by the ancient Greeks, and here it is. It's probably my favorite proof in this book.

Theorem 4.9 $\sqrt{2}$ *is irrational.*

Proof   This will be a proof by contradiction. Suppose there are integers a and b, $b \neq 0$, so that $\frac{a}{b} = \sqrt{2}$. If b is negative, then so must a be. In this case negate both a and b, so that both are now positive, calling the results still a and b. Furthermore, we may convert the fraction $\frac{a}{b}$ to lowest terms by repeatedly dividing out by common factors. Eventually the final fraction will be in lowest terms. Let $\frac{a}{b}$ be the final fraction. We know that a and b can't *both* be even (though maybe exactly one is), for then $\frac{a}{b}$ wouldn't be in lowest terms. So we have:

$$\frac{a}{b} = \sqrt{2},$$

with a, b not both even. If we square both sides of this equation, we get:

$$\frac{a^2}{b^2} = 2,$$

implying that

$$a^2 = 2b^2.$$

The right-hand side is 2 times b^2, and b^2 is an integer (because b itself is an integer). Hence the right-hand side is even, as it is 2 times an integer.

The left-hand side is a^2, which, equaling the even number $2b^2$, must be even. Now here comes the tricky part. Not only is a^2 even, but so is a itself. For example, if $a^2 = 64$ is even, then a itself (which in this case equals 8) is even. If $a^2 = 144$ is even, then so is a (which in this case is 12). Now this point is really not obvious—it wasn't obvious to me when I first saw the proof. Let's prove this simple fact, while still amidst the proof of the theorem. ∎

Lemma 4.1 *If a is an integer and a^2 is a multiple of 2 (i.e., is even), then a itself (not just a^2) is a multiple of 2 (i.e., is even).*

Proof There are only two possibilities for a: that it is even, and that it is odd. We will prove that a cannot be odd, which will prove that a is even. Suppose a were odd. This means that if one divides a by 2, the quotient would be some integer k, and the remainder would be 1. In other words, $a = 2k + 1$. But this means that $a^2 = (2k+1)(2k+1)$. If you multiply out all four terms in $(2k+1)(2k+1)$, you'll get $4k^2 + 2k + 2k + 1 = 4k^2 + 4k + 1$. Now $4k^2 + 4k + 1 = 2(2k^2 + 2k) + 1$. But $2k^2 + 2k$ is an integer, so this number is twice an integer plus 1, so is odd. We have shown that if a is odd, then so is a^2. But we were told that a^2 was *even*, so that means that a cannot be odd, and hence must be even. We have finished the proof of the lemma. ∎

Now let's continue the main proof. $a^2 = 2b^2$, so a^2 is even. By the lemma, a is itself even. That means there is some integer r such that $a = 2r$. Hence $a^2 = (2r)^2 = 4r^2$. But $a^2 = 2b^2$, so $4r^2 = 2b^2$ and hence, dividing both sides by 2, $2r^2 = b^2$.

We now know that b^2, being twice the integer r^2, is even. Applying that handy-dandy lemma again, since b^2 is even, b *itself* must be even.

But wait. We said that a and b aren't *both* even. But they are both even! This is the contradiction, which arose solely because we assumed that $\sqrt{2}$ is rational. It must be the case that $\sqrt{2}$ is irrational. ∎

You have to understand how momentous the discovery of the irrationality of $\sqrt{2}$ was. Until $\sqrt{2}$ was proven irrational, around the year -450 (also known as 450 BC or BCE), everyone thought that all numbers were rational. The square root of two was the first number proven irrational in history! The proof is credited to the

Greek school of mathematics founded by Pythagoras, however, who from the school proved the irrationality has been lost to history.

The proof is sometimes credited, perhaps to his dismay, to Hippasus of Metapontum, who died approximately in the year -450. I write "to his dismay" because of the colorful stories about the discoverer of irrational numbers. Some say he drowned, as a punishment from the gods. Others say that he announced his discovery on board a ship and that the discovery so traumatized his shipmates that they threw him overboard to drown. Others say that Pythagoras himself sentenced the proof's author to drown for proving that $\sqrt{2}$ is irrational. If that last story is true, then school administrators really had way too much power in those days.

Now we will prove an even more interesting theorem, namely, that the square root of every positive integer x is irrational, unless $\sqrt{x}$ is an integer. (Notice that I am not saying that $\sqrt{x}$ for a positive integer x is irrational unless $\sqrt{x}$ is rational, which would be a tautology,[2] but instead that $\sqrt{x}$ is irrational unless $\sqrt{x}$ is an *integer*, which is very nontrivial.)

Theorem 4.10 *Let x be a positive integer such that $\sqrt{x}$ is not an integer. Then $\sqrt{x}$ is not rational.*

This theorem subsumes the theorem that $\sqrt{2}$ is irrational, since $\sqrt{2}$, being strictly between 1 and 2, is not an integer. This theorem implies that $\sqrt{2}, \sqrt{3}, \sqrt{5}, \sqrt{6}, \sqrt{7}, \sqrt{8}, \sqrt{10}, \ldots$ are irrational. (Clearly $\sqrt{4}$ and $\sqrt{9}$, being integers, are rational.)

Proof Let $z = \sqrt{x} > 0$. Suppose for a contradiction that z is rational, so that there are positive integers a, b such that $z = a/b$ and a and b have no divisors in common, that is, a/b is a fraction in lowest terms. We know from Theorem 1.1 that any positive integer can be written as a product of primes. Let us write $a = p_1 p_2 p_3 \cdots p_m$ and $b = q_1 q_2 q_3 \cdots q_n$ where the p_i's and q_j's are primes.

We have to be a bit careful. Either n, the number of q_j's, is positive or it's zero. If it's zero, then $b = q_1 q_2 \cdots q_n$ is an empty product, which is defined to be 1, which would mean that $z = a/1$ where a is an integer, and hence $\sqrt{x}$ is the integer $a/1$, which we know it is not. Hence we know that $n \geq 1$. This means that we can talk about q_1. (It wouldn't have made any sense to talk about q_1 if n had been 0.)

We know, since a and b have no divisors in common, that clearly q_1, which is a divisor of b, is not a divisor of a. Hence no p_i equals q_1, for otherwise a and b would have the divisor q_1 is common. Remember this point.

We have

$$z = \frac{a}{b} = \frac{p_1 p_2 \cdots p_m}{q_1 q_2 \cdots q_n}$$

and hence

---

[2] (Like the one I tell my kids: "I never lie to you except when I do").

$$x = z^2 = \frac{a^2}{b^2} = \frac{p_1^2 p_2^2 \cdots p_m^2}{q_1^2 q_2^2 \cdots q_n^2}.$$

Since x is an integer,

$$x \cdot q_1^2 q_2^2 \cdots q_n^2 = p_1 p_1 p_2 p_2 \cdots p_m p_m.$$

The left-hand side of this equation is an integer which is clearly divisible by q_1. By Corollary 2.2, since q_1 is a prime dividing the product $p_1 p_1 p_2 p_2 \cdots p_m p_m$ of primes, q_1 must equal either p_1, p_1, p_2, p_2, ..., p_m or p_m. In other words, some p_i equals q_1.

However, we said above that no p_i equals q_1! Hence we have found a contradiction to the assumption that $z = \sqrt{x}$ is rational, and hence we have proven that $\sqrt{x}$ is irrational. ■

4.4   Puzzle

Let n (e.g., 2023) be the current year as a positive integer. Write n as a sum of positive integers such that the product of those integers is maximized. For example, $2023 = 1000 + 1000 + 23$, having product 23000000, but this is not the maximum achievable.

4.5   Exercises

Exercise 4.1 An *implication* is a statement of the form, "If something is true, then something else is true," or "If P, then Q," or "$P \Rightarrow Q$" (read "P implies Q"). For example, "If x is a multiple of 3, then x^2 is a multiple of 3," which is true.

(a) Prove this statement. Use the fact that x is a multiple of 3 if and only if there is an integer k such that $x = 3k$.

   The *converse* of an implication "If P, then Q," is the implication "If Q, then P." It is a very important fact that even if the first implication is true, the second one (the converse) may be false.

(b) Prove that the implication, "If x is a multiple of 4, then x^2 is a multiple of 4," is true.

(c) The converse of this implication is, "If x^2 is a multiple of 4, then x is a multiple of 4." Prove that this converse is false. To prove the converse is false, it is enough to give *one* x for which the conclusion fails.

Exercise 4.2 The converse of "If x is a multiple of 3, then x^2 is a multiple of 3," which you proved to be true in Exercise 4.1, is this statement: "If x^2 is a multiple of 3, then x is a multiple of 3." Prove that this converse is true.

Exercise 4.3 Use the result of the previous problem to prove directly, as in the proof of Theorem 4.9, that $\sqrt{3}$ is irrational.

Exercise 4.4 One would like to prove, using the method of the previous problem, that for any prime p, $\sqrt{p}$ is irrational. This is certainly true, since it is implied by Theorem 4.10. However, it is harder to prove this than to prove that $\sqrt{2}$ and $\sqrt{3}$ are irrational. Explain in words why the proofs that $\sqrt{2}$ and $\sqrt{3}$ are irrational don't immediately show that $\sqrt{p}$ is irrational for every p.

Exercise 4.5 A *continued fraction* is a finite or infinite representation of a nonnegative real α as

$$\alpha = a_1 + \cfrac{1}{a_2 + \cfrac{1}{a_3 + \cfrac{1}{a_4 + \cfrac{1}{a_5 + \cfrac{1}{a_6 + \cdots}}}}}$$

in which each a_i, known as a *partial quotient*, is a positive integer, except that a_1 is allowed to be 0. Let me give an example:

$$\frac{67}{20} = 3 + \frac{7}{20}$$
$$= 3 + \cfrac{1}{\frac{20}{7}}$$
$$= 3 + \cfrac{1}{2 + \frac{6}{7}}$$
$$= 3 + \cfrac{1}{2 + \cfrac{1}{\frac{7}{6}}}$$
$$= 3 + \cfrac{1}{2 + \cfrac{1}{1 + \frac{1}{6}}}.$$

The partial quotients here are 3,2,1,6. All the "numerators" in a continued fraction are required to be 1.

You can think of this iteratively. You start by looking for the continued fraction representation of 67/20. It's 3 plus 1 over (the continued fraction expansion of 20/7). That is 2 plus 1 over (the continued fraction expansion of 7/6). That is 1 plus 1/6. Done. Effectively you replaced the pair (67, 20) by the pair (20, 7), which you then replaced by (7, 6).

The natural algorithm suggested by this example is very simple. Algorithm cf_expansion(α) prints the partial quotients of the continued fraction expansion of nonnegative real α. The *floor* $\lfloor x \rfloor$ of a real x is the largest integer less than or equal to x.

Algorithm cf_expansion(α):

1. Let $q = \lfloor \alpha \rfloor$
2. Print q
3. If $q \neq \alpha$, return cf_expansion($1/(\alpha - q)$)

This algorithm requires exact arithmetic on reals, which no computer supplies, so we will not run it on a real computer. It is not obvious that this algorithm terminates, even if α is rational, though we will see that it does. If α is irrational, it will run forever.

By hand, find the continued fraction representation of $531/187$.

Exercise 4.6 (Continuation)
Compare the example in Exercise 4.5 of finding the continued fraction expansion of $67/20$ to a run, shown here, of the Euclidean Algorithm on the pair $(67, 20)$:

$$67 = 3 \cdot 20 + 7$$
$$20 = 2 \cdot 7 + 6$$
$$7 = 1 \cdot 6 + 1$$
$$6 = 6 \cdot 1 + 0.$$

For $\beta \geq 0$, let $CF(\beta)$ be the continued fraction representation of β. Suppose α is a real at least one. The continued fraction algorithm sets $q = \lfloor \alpha \rfloor$. If $\alpha = q$, the algorithm stops there with $CF(\alpha) = q$. Otherwise, it sets

$$CF(\alpha) = q + \frac{1}{CF\left(\frac{1}{\alpha - q}\right)}.$$

As mentioned in Exercise 4.5, it is not obvious this algorithm terminates, even for rational α.

Now let's examine the special case in which α is a rational; specifically, let us suppose $\alpha = b/a \geq 1$ where $0 < a \leq b$ are relatively prime integers. Unless α is an integer, we have:

$$CF(\alpha) = q + \frac{1}{CF\left(\frac{1}{\alpha - q}\right)},$$

where $q = \lfloor b/a \rfloor$ and $1/(\alpha - q) = 1/(b/a - q) = a/(b - aq)$. So the numerator-denominator pair (b, a) has been replaced by the numerator-denominator pair $(a, b - aq)$. Compare this to the Euclidean Algorithm run on the pair (b, a): in the first step, the algorithm writes $b = qa + r$ for the *same* q (and for $r = b - aq$), and then it replaces (b, a) by $(a, r) = (a, b - aq)$. This is exactly the same transformation made in the continued fraction algorithm! While not a rigorous proof –an inductive proof would be rigorous– this argument elucidates the tight connection between the continued fraction representation of a rational $b/a \geq 1$ in lowest terms and a run of the Euclidean Algorithm on the pair (b, a).

(a) Prove that any positive rational number has a finite continued fraction representation.
(b) Prove that any positive real with a finite continued fraction representation is rational.

   Hence the continued fraction expansion of a positive rational α is finite if and only if α is rational.

Exercise 4.7 * See Exercise 4.5 for the definition of continued fraction:

(a) What is the exact value of this infinite continued fraction:

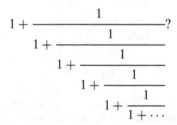

   Do not use a calculator or computer.
(b) Find the continued fraction expansion of $1 + \sqrt{2}$.

Exercise 4.8

(a) Write a Python program to take a description of a finite continued fraction and compute its value.
(b) Here is the infinite continued fraction representation of a number I'll call ω, the last letter of the Greek alphabet, called "omega":

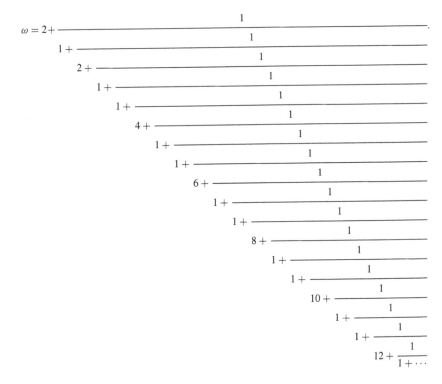

$$\omega = 2 + \cfrac{1}{1 + \cfrac{1}{2 + \cfrac{1}{1 + \cfrac{1}{1 + \cfrac{1}{4 + \cfrac{1}{1 + \cfrac{1}{1 + \cfrac{1}{6 + \cfrac{1}{1 + \cfrac{1}{1 + \cfrac{1}{8 + \cfrac{1}{1 + \cfrac{1}{1 + \cfrac{1}{10 + \cfrac{1}{1 + \cfrac{1}{1 + \cfrac{1}{12 + \cfrac{1}{1 + \cdots}}}}}}}}}}}}}}}}}.$$

(Aside from the anomalous first 2, do you see the pattern?)

Run your program from part (a) to estimate the value of ω. What is the standard name for ω?

(c) Prove that ω is irrational.

Exercise 4.9 Suppose that $n \geq 2$. Let α be the (positive) nth root of 2. Prove that α is irrational.

Exercise 4.10 * Prove there exists a pair (a, b) of irrationals such that a^b is rational.

Exercise 4.11 Prove Theorem 4.3.

Exercise 4.12 True or false? $\mathbb{Q}$ denotes the set of rationals.

(a) If $x \in \mathbb{Q}$, then $x/2 \in \mathbb{Q}$.
(b) If $x \notin \mathbb{Q}$, then $x/2 \notin \mathbb{Q}$.
(c) If $x, y \in \mathbb{Q}$, then $x + y \in \mathbb{Q}$.
(d) If $x, y \notin \mathbb{Q}$, then $x + y \notin \mathbb{Q}$.
(e) If $x \in \mathbb{Q}$, then $x^2 \in \mathbb{Q}$.
(f) If $x \notin \mathbb{Q}$, then $x^2 \notin \mathbb{Q}$.
(g) If $x \in \mathbb{Q}$, $x > 0$, then $\sqrt{x} \in \mathbb{Q}$.

(h) If $x \notin \mathbb{Q}$, $x > 0$, then $\sqrt{x} \notin \mathbb{Q}$.

(i) If $x \in \mathbb{Q}$, $y \notin \mathbb{Q}$, then $x + y \notin \mathbb{Q}$.

(j) If $x, y \in \mathbb{Q}$, $x, y > 0$, then $x^y \notin \mathbb{Q}$.

(k) If $x, y \notin \mathbb{Q}$, $x, y > 0$, then $x^y \notin \mathbb{Q}$.

Exercise 4.13 In the Euclidean plane, points are given by pairs of real numbers. Say we have two points $P_1 = (x_1, y_1)$ and $P_2 = (x_2, y_2)$. Let $\Delta_1 = |x_2 - x_1|$ and $\Delta_2 = |y_2 - y_1|$. Here $|z|$, the *absolute value* of z, is z or $-z$, whichever is nonnegative, and the Δ is a capital Greek letter "Delta." We usually define the distance between P_1 and P_2 to be $\sqrt{\Delta_1^2 + \Delta_2^2}$, known as the *Euclidean distance* and denoted $d_2(P_1, P_2)$ because of the "2" in the exponent, but this is not the only way to define distance. There are also

$$d_\infty(P_1, P_2) = \max\{\Delta_1, \Delta_2\}$$

and

$$d_1(P_1, P_2) = \Delta_1 + \Delta_2,$$

among others. (The "1" subscript on the d refers to the powers of 1 of Δ_1 and Δ_2 which are being used.) This last one is often called the *Manhattan metric*, because it is the distance along the grid lines defined by streets in Manhattan, New York (where I live, in fact).

(a) Show that for any P_1, P_2, $d_\infty(P_1, P_2) \le d_2(P_1, P_2)$.

(b) Show that for any P_1, P_2, $d_2(P_1, P_2) \le d_1(P_1, P_2)$.

   Hence:

$$d_\infty(P_1, P_2) \le d_2(P_1, P_2) \le d_1(P_1, P_2).$$

(We in Manhattan are stuck with the most costly metric of the three.)

Exercise 4.14 * Show that $\log_2 10$ is irrational.

Exercise 4.15 Let p, q be primes. Show that $\sqrt{p} + \sqrt{q}$ is irrational.

4.5.1   Hints

4.7 Write an equation for the number in terms of itself and solve it.

4.10 Show that either (a, b) equaling $(\sqrt{2}, \sqrt{2})$ or $(\sqrt{2}^{\sqrt{2}}, \sqrt{2})$ works.

4.14 Look at the prime divisors of 2 and 10.

5.1   Roots of Polynomials

You probably have seen this definition before.

Definition 5.1 A *polynomial* is a finite sum of terms, each of which is a *coefficient* multiplied by a variable raised to a nonnegative integral power. In other words, it is a sum

$$c_0 + c_1 x + c_2 x^2 + c_3 x^3 + \cdots + c_n x^n;$$

if $c_n \neq 0$, its *degree* is n. A *root* of a polynomial is a number that, when plugged in as the variable, causes the polynomial to evaluate to 0.

Example -17, $-4 + 9x^2$, $-4y - 23y^3 + \pi y^5$ are all polynomials. $x = 2/3$ and $x = -2/3$ are roots of $-4 + 9x^2$.

As we saw in Chap. 4, negative numbers were introduced approximately 2000 years ago. The introduction of negative numbers allowed one to solve any "linear" equation. A *linear equation*, which comes from the formula $y = ax + b$ for a line (and hence the name "linear"), is any equation of the form $ax + b = 0$ where a and b are reals and a is not zero. We require a to be nonzero so as to disallow unpleasant equations like $0 \cdot x + 3 = 5$, which has no solution. For example, $4x + 3 = 0$ is a linear equation.

Without negative numbers, you couldn't solve all linear equations with positive coefficients. For example, you couldn't even solve $x + 1 = 0$. *With* negative numbers, solving linear equations is a snap, even if a and b are themselves allowed to be negative: $a \cdot x + b = 0$ is true if, and only if, $a \cdot x = -b$ (by subtracting b from both sides). Since a is assumed to be nonzero, we can divide both sides by a, and we

H. Karloff, *Mathematical Thinking*, Compact Textbooks in Mathematics,
https://doi.org/10.1007/978-3-031-33203-6_5

infer that $x = (-b)/a$. For example, $3x + 2 = 0$ has solution $x = (-2)/3 = -2/3$. Notice that there is exactly one solution to the equation $ax + b = 0$ if $a \neq 0$.

What happens when you want to solve an equation with a quadratic term in x (one involving x^2), such as $x^2 - 4x + 3 = 0$? This particular quadratic equation is easy to solve. It so happens (and I chose it this way) that $x^2 - 4x + 3 = (x-1)(x-3)$, so $x^2 - 4x + 3 = 0$ if and only if $(x - 1)(x - 3) = 0$. Since the product of reals can equal 0 if and only if at least one of the factors is 0, we know that $(x - 1)(x - 3) = 0$ if and only if $x - 1 = 0$, that is, $x = 1$, or $x - 3 = 0$, that is, $x = 3$.

(By the way, you may remember from Chap. 3 on modular arithmetic that when one does arithmetic in the integers mod 4, you *can* get 0 by multiplying together nonzeros 2 and 2, since $2 \cdot 2 = 4 \equiv 0 \pmod 4$. This cannot happen in the reals, or in the integers mod a prime modulus.)

Let's get back to quadratic equations. Consider an equation $ax^2 + bx + c = 0$, where we may assume that $a \neq 0$, since if $a = 0$ we have a linear equation which we know how to solve. You have probably already learned how to solve a quadratic equation in school, but in any case, I will show you how now. Feel free to skip this part if you've learned how to solve quadratic equations already. Consider the equation $ax^2 + bx + c = 0$. Since $a \neq 0$, we can divide both sides by a without changing the solutions: $ax^2 + bx + c = 0$ if and only if

$$x^2 + \frac{b}{a}x + \frac{c}{a} = 0. \tag{5.1}$$

The trick now is to get rid of that pesky linear term $(b/a)x$. Notice that $(x + b/(2a))^2 = (x + b/(2a))(x + b/(2a))$, which we evaluate by taking all four terms $x \cdot x$, $x \cdot (b/(2a))$, $(b/(2a)) \cdot x$, and, last, $(b/(2a)) \cdot (b/(2a))$. If you add these four terms together, you get $x^2 + x \cdot (b/a) + b^2/(4a^2)$. Notice the similarity to Eq. (5.1). We have the x^2 and the $(b/a)x$ we wanted, but we're missing the c/a and we have a superfluous $b^2/(4a^2)$. We can correct for these "mistakes" by adding c/a and subtracting $b^2/(4a^2)$. That means that

$$x^2 + \frac{b}{a}x + \frac{c}{a} = \left(x + \frac{b}{2a}\right)^2 + \frac{c}{a} - \frac{b^2}{4a^2},$$

and hence

$$x^2 + \frac{b}{a}x + \frac{c}{a} = 0$$

if and only if

$$\left(x + \frac{b}{2a}\right)^2 = \frac{b^2}{4a^2} - \frac{c}{a} = \frac{b^2 - 4ac}{4a^2}.$$

Remember that $\sqrt{z}$ is the *nonnegative* number whose square is z (if such a number exists, which happens if and only if z is nonnegative itself). Now let $y = x + b/(2a)$. We have

$$y^2 = \frac{b^2 - 4ac}{4a^2}.$$

The denominator $4a^2$ is positive. If $b^2 - 4ac \geq 0$, then $(b^2 - 4ac)/(4a^2) \geq 0$, and we can take its square root. Since $y^2 = z$ has solutions $y = \sqrt{z}$ and $y = -\sqrt{z}$, we infer that $y = \sqrt{b^2 - 4ac}/\sqrt{4a^2}$ or $y = -\sqrt{b^2 - 4ac}/\sqrt{4a^2}$. What is $\sqrt{4a^2}$? It's tempting to say $2a$, since $(2a)^2 = 4a^2$, but remember that $\sqrt{z}$ must be nonnegative, so $\sqrt{4a^2} = 2|a|$, where $|a|$, the *absolute value* of a, is either a or $-a$, whichever is nonnegative. So $y = \pm\sqrt{b^2 - 4ac}/(2|a|)$. Here the "$\pm$," which means "plus or minus," saves us: since we already have $\pm$ at the front, we don't need the absolute value sign around the a in the denominator after all.

Since $y = \pm\sqrt{b^2 - 4ac}/(2a)$ and $y = x + b/(2a)$, we infer that

$$x = \frac{-b \pm \sqrt{b^2 - 4ac}}{2a}.$$

Whew! That was a lot of work. But now the really interesting part begins. What happens if $b^2 - 4ac$ is *negative*, so we can't take its square root? One answer is, well, there's no real solution, and that is true. But that's somewhat like saying that $x + 1 = 0$ has no solution among the nonnegative reals, which is also true, but which is a problem which is easily remedied by adding the negative numbers. Maybe we should do something similar here?

Let's take the simplest example of a quadratic equation with no real solution: $x^2 = -1$. This really has no real solution, since the square of any real x is nonnegative, so it can't be -1. So the inescapable conclusion is that there's no real solution.

It seems like such a shame. Just as we concocted -1 to be a solution to $x + 1 = 0$, let's magically concoct a solution to the equation $x^2 = -1$. In fact, since we're just imagining there's a solution, let's call the solution i for "imaginary." So now we have a solution to $x^2 = -1$. In fact, we have two, since if $i^2 = -1$, then also $(-i)^2 = (-i)(-i) = (-1 \cdot -1)(i \cdot i) = 1(-1) = -1$; hence the two solutions are i and $-i$.

We cheated, no? Where did that i come from? We really said, "There's no real solution, so let's make up a nonreal one." Really, that's what we did. I remember when I saw i for the first time, I recoiled in horror. No! There's no number whose square is -1. No! I couldn't accept this mathematical sleight-of-hand. It took me quite a while to get used to i.

Now just as when we added -1 to the nonnegatives, we also added -2, -3, $-1/2$, etc., we have to add some more numbers here. What happens when you add 3 to i? Well, you get $3 + i$. That's how you write it—you can't simplify it. When you add -4 to i, you get $-4 + i$. What happens when you multiply i by 2? Well, you get $2i$. What if you multiply i by 2 and then add 3? You get $3 + 2i$. What happens when you add $3 + 4i$ to $5 + 6i$? You get $(3 + 5) + (4 + 6)i = 8 + 10i$. If you want the general formula, it's $(a + bi) + (c + di) = (a + c) + (b + d)i$, if a, b, c, d are reals.

Subtraction is easy: $(a + bi) - (c + di) = (a - c) + (b - d)i$.

How about multiplication? It's not hard. $(3 + 4i)(5 + 6i) = 3 \cdot 5 + 3 \cdot (6i) + (4i) \cdot 5 + (4 \cdot 6)(i \cdot i) = (15 - 24) + i(18 + 20) = -9 + 38i$. In general, $(a + bi)(c + di) = (ac - bd) + (ad + bc)i$.

The only tricky one is division. Let's say we want $(3 + 4i)/(5 + 6i)$. We're not allowed to have i's in denominators here, so we "clear out" the i in the denominator by multiplying both $3 + 4i$ (the numerator) and $5 + 6i$ (the denominator) by $5 - 6i$, which is called the "conjugate" of $5 + 6i$. Notice that $(5 + 6i)(5 - 6i) = 25 - 30i + 30i - 36(i^2) = 25 + 36 = 61$; the i has disappeared!

$$
\begin{aligned}
\frac{3 + 4i}{5 + 6i} &= \frac{(3 + 4i)(5 - 6i)}{(5 + 6i)(5 - 6i)} \\[2mm]
&= \frac{15 - 18i + 20i - 24(-1)}{61} \\[2mm]
&= \frac{39 + 2i}{61} \\[2mm]
&= \frac{39}{61} + \frac{2}{61}i.
\end{aligned}
$$

The same trick works whatever nonzero number goes in the denominator. If it's $c + di$, then multiply both the numerator and denominator by $c - di$. The general rule is

$$
\begin{aligned}
\frac{a + bi}{c + di} &= \frac{(a + bi)(c - di)}{(c + di)(c - di)} \\[2mm]
&= \frac{(ac + bd) + (-ad + bc)i}{c^2 + d^2} \\[2mm]
&= \frac{ac + bd}{c^2 + d^2} + \frac{-ad + bc}{c^2 + d^2}i.
\end{aligned}
$$

Note that if $c + di$ is not 0, then either c or d is nonzero, in which case the denominator $c^2 + d^2$ is positive.

These numbers $a + bi$, with a and b real, are called *complex numbers*, and, to my great surprise, are heavily used in engineering and physics, and even in more advanced number theory than the kind we've been doing in this book. Who would've thought adding i would be so fruitful?

Here's what's really interesting about the complex numbers. We started by adding i because we wanted $x^2 + 1 = 0$ –just *one* small and simple equation– to have a solution. Then we added numbers of the form $a + bi$. With these numbers, we can find solutions to all quadratic equations. But *what about cubic equations*, those equations like $4x^3 - 2x^2 + 17x - 14 = 0$ which have an x^3 term? What *else* does one have to add, on top of the $a + bi$'s we've already added? And after adding those numbers, what do we have to add to find solutions to *quartic equations*, those equations like $3x^4 - 2x^3 - x + 83 = 0$? And how about quintics, those with x^5 terms? How long will this go on? How many new "numbers" must we add?

The amazing fact is that the answer is *none*. Once you add the $a + bi$'s, there's nothing more to add: every cubic (degree-3) polynomial equation has exactly three solutions (although they need not all be different); every quartic (degree-4) polynomial equation has exactly four solutions; every quintic (degree-5) polynomial equation has exactly five solutions; and so on.

But wait, there's more. (I sound a bit like a late-night TV salesman.) You can not only find solutions to all of the polynomial equations with real coefficients, but you can also find solutions even to those with *complex* coefficients (and an equation of degree d has exactly d solutions). This is a remarkable fact known as the Fundamental Theorem of Algebra.

Theorem 5.1 *The Fundamental Theorem of Algebra. If $p(x)$ is a polynomial equation of degree $d \geq 1$ with complex coefficients, then $p(x)$ has exactly d complex roots (which need not all be different).*

Unfortunately the proof is too complicated to include in this book.

It is astonishing to me that by adding the $a + bi$'s to get solutions to quadratic equations, we got solutions for free to all higher-degree equations as well, and even to those with complex coefficients. This is a point I certainly didn't appreciate when I learned it.

By the way, you may be asking yourself, "What does he mean by having d roots which need not all be different? Why count some roots more than once?" What *do* I mean by saying that some roots count more than once? The answer is that a polynomial equation

$$p(x) = 0,$$

in which

$$p(x) = a_d x^d + a_{d-1} x^{d-1} + a_{d-2} x^{d-2} + \cdots + a_1 x + a_0$$

and in which the a's are complex, can always be written as $a_d (x - r_1)(x - r_2)(x - r_3) \cdots (x - r_d)$, where the r's are complex. Each r_j is a root of the polynomial, since if $x = r_j$, clearly $p(x)$ becomes 0. Altogether there are d roots. The *multiplicity* of a root s is the number of j's such that $s = r_j$. For example, $x^3 - x^2 - 8x + 12 = (x - 2)(x - 2)(x + 3)$, which you can verify yourself, meaning that 2 is a double root and -3 is a single root.

For quadratic equations $p(x) = ax^2 + bx + c = 0$, we derived a nice formula:

$$x = \frac{-b \pm \sqrt{b^2 - 4ac}}{2a}.$$

Are there similar formulas for the roots of higher-degree equations?

For cubic polynomial equations, yes, there is a formula. It involves a finite number of addition, subtraction, multiplication, division, and root-extraction steps. It's complicated so I won't give it here, but you can find it on the web if you want to see it.

How about quartics? Again, there is a formula. This one is even more complicated, but evaluating it still involves a finite number of additions, subtractions, multiplications, divisions, and root extractions.

And how about quintics? Here the answer is "no." It is not "no," as in "no one has found such a formula yet," it is "no, no formula can possibly exist, as proven by Abel in 1826." This is really fascinating. In fact, the roots of even the simple polynomial $x^5 - x - 1 = 0$ can't be found with a finite number of additions, subtractions, multiplications, divisions, and root extractions.

5.2 The Geometric Representation and Norm of a Complex Number

This is almost too simple: the geometric representation of the complex number $a+bi$ is the point (a, b) in the usual Euclidean plane. That is, $2 + 3i$ is represented by the point $(2, 3)$, and $-4 - 3i$ is represented by the point $(-4, -3)$.

The norm is more interesting.

Definition 5.2 The *norm*, or *absolute value*, or *modulus* of a complex number $z = a + bi$ is written $|z|$ and is defined to be $\sqrt{a^2 + b^2}$, which is always a nonnegative real number.

For example, the norm $|3 - 4i|$ of $3 - 4i$ is $\sqrt{9 + 16} = 5$.

You may be asking yourself, since the absolute value of a real number x is denoted $|x|$, how do I know, when I see those vertical bars, whether I'm talking about absolute value of a real number or norm of a complex number? This is confusing, is it not, as all real numbers a can also be viewed as complex numbers $a + 0i$? The good news is that it doesn't matter. If the complex number $z = a + bi$ happens to be real, then its norm is exactly the absolute value of a. This is true because when $b = 0$, $\sqrt{a^2 + b^2} = \sqrt{a^2}$, which is a if $a \geq 0$ and $-a$ if $a < 0$, coinciding with absolute value of a.

By the way, the norm $|z|$ of a complex number $z = a+bi$ is exactly the Euclidean distance from the geometric representation (a, b) of z to the origin $(0, 0)$.

Since writing square root so often gets to be tedious, one often works with the *square* of the norm of z, written $|z|^2$.

The norm has some fairly magical properties. The norm of a product is the product of the norms.

Theorem 5.2 *For complex numbers z_1 and z_2, $|z_1 \cdot z_2| = |z_1||z_2|$.*

Proof Suppose $z_1 = a + bi$ and $z_2 = c + di$.

$$(a + bi)(c + di) = (ac - bd) + i(ad + bc).$$

Hence the square of its norm is

$$(ac-bd)^2+(ad+bc)^2 = [(ac)^2+(bd)^2-2(ac)(bd)]+[(ad)^2+(bc)^2+2(ad)(bc)].$$

The $-2(ac)(bd)$ cancels with the $2(ad)(bc)$, leaving us

$$(ac)^2 + (bd)^2 + (ad)^2 + (bc)^2.$$

On the other hand, $|z_1|^2|z_2|^2$ equals

$$(a^2 + b^2)(c^2 + d^2) = a^2c^2 + a^2d^2 + b^2c^2 + b^2d^2,$$

which is the same thing. This means that

$$|z_1z_2|^2 = |z_1|^2|z_2|^2,$$

and, taking square roots of both sides, we get:

$$|z_1z_2| = |z_1||z_2|. \qquad \blacksquare$$

Theorem 5.3 *For complex numbers z_1 and z_2, $|z_1 + z_2| \leq |z_1| + |z_2|$.*

Proving this theorem is Exercise 5.1.

5.3   *e* **to a Complex Power**

Remember the famous and important number e, which is approximately 2.718281828459, from Chap. 1? Now e is a real number, so if a is a real number, then e^a, the ath power of e, makes perfect sense, whether a is positive, negative, or 0. For example, we already know that $e^{2.3}$ is approximately 10. Furthermore, e^x is always positive for a real x; it cannot be negative and it cannot be 0. But what is e^z if z is *complex*? It turns out that there is a remarkable connection between the exponential function e^x and trigonometry, triggered by the question, "What is e^z if z is a *complex* number?"

Let z be a complex number. Normally we would write $z = a + bi$, where a and b are real, but instead let's write $z = x + \theta i$, where x and θ are real. (Why that Greek letter θ, a "theta?" Because θ is often used to represent an angle, and we are shooting for a connection with trigonometry.) The question is, what is $e^{x+\theta i}$? Since we'd like an exponential of a sum to be the product of the exponentials, it had better be true that $e^{x+\theta i} = e^x \cdot e^{\theta i}$. Since x is a real, we know what e^x is. It turns out that there's a remarkable expression for $e^{\theta i}$ connecting it to trigonometry. Personally, I'm still surprised that there's a relationship between the exponential function and trigonometry.

I'll give the expression first and explain where it comes from in a moment. It uses the sine and cosine functions of trigonometry.

Theorem 5.4 *For any real number* θ, *$e^{\theta i} = \cos \theta + i \cdot \sin \theta$.*

Here sin and cos are respectively the sine and cosine functions from trigonometry –I'll define them below– where the angle is given in radians. An angle of π radians is $180°$ –a straight line– and hence $\pi/2$ radians is $90°$, a right angle, and $\pi/4$ radians is $45°$. Also $\pi/6$ is $30°$ and $\pi/3$ is $60°$.

You're probably thinking, "Why do we have to use those annoying radians, which involve π's, instead of just normal degrees, which are so much simpler?" Fair question. The answer is that one could use degrees instead, but then many of the formulas would have lots of messy 180's and π's in them. With radians, they have neither.

In case you've forgotten trigonometry or never learned it, let me define sine and cosine. Draw a unit circle centered at the origin. Let θ be any real number. Starting from the point $(1, 0)$ on the positive x-axis and on the unit circle, go around counterclockwise on the circle θ radians, stopping at, say, (a, b) on the unit circle; define $\cos(\theta) = a$ and $\sin(\theta) = b$.

For example, if $\theta = \pi/2$ (a right angle), you will stop at $(0, 1)$ on the positive y-axis; hence the cosine of $\pi/2$ is 0 and the sine of $\pi/2$ is 1. Since taking $\theta = 0$ takes us nowhere, and we started at $(1, 0)$, $\cos 0 = 1$ and $\sin 0 = 0$. The Pythagorean Theorem tells us that taking $\theta = \pi/4$ will take us to $(\sqrt{2}/2, \sqrt{2}/2)$, and hence $\cos(\pi/4) = \sin(\pi/4) = \sqrt{2}/2 \approx 0.707107$. If n is any integer (positive, negative, or zero) and $\theta = 2\pi n$, you will finish at $a = 1$ and $b = 0$ (precisely where you started), so $\cos(2\pi n) = 1$ and $\sin(2\pi n) = 0$. (A negative angle θ necessitates movement clockwise by $-\theta$ radians.) Last, taking $\theta = \pi$ will take us to $(-1, 0)$; hence $\cos \pi = -1$ and $\sin \pi = 0$. Plugging these last values into Theorem 5.4 gives us a famous formula tying together five very important numbers, 0, 1, e, i, and π.

Corollary 5.1 *Euler's Identity.* $e^{\pi i} + 1 = 0$.

Proof Plug in $\theta = \pi$ in Theorem 5.4. We get $e^{\pi i} = -1 + 0i$.   ∎

Uniting the fundamental constants $0, 1, \pi, e$, and i together with the basic operations addition, multiplication, and exponentiation, Euler's Identity is considered by many to be the most beautiful theorem in all of mathematics.

5.4 Infinite Series

I hope you're asking yourself where Theorem 5.4 came from. How is e^z defined for a complex number z?

It turns out that, using calculus, one can write down an *infinite* series, called a *Taylor series*, for e^x when x is a real number. Remember that $n!$, read "n factorial," is just $1 \cdot 2 \cdot 3 \cdots n$. For example, $4! = 1 \cdot 2 \cdot 3 \cdot 4 = 24$. Here is the Taylor series for e^x, where x is a real number:

$$e^x = 1 + \frac{x}{1!} + \frac{x^2}{2!} + \frac{x^3}{3!} + \frac{x^4}{4!} + \cdots .$$

Don't worry about the fact that the series goes on forever. The denominators grow faster than the numerators so for any x, even for a large one, the infinite series converges to a finite number. For example, since $e^1 = e \approx 2.718$, if you plug in $x = 1$ in the Taylor series, you get

$$e^1 = 1 + \frac{1}{1!} + \frac{1}{2!} + \frac{1}{3!} + \frac{1}{4!} + \cdots ,$$

Even the sum shown, up to and including the term $\frac{1}{4!}$, gives $2.7083333\ldots$, which is a decent approximation to e.

By the way, one can construct Taylor series for $\cos x$ and $\sin x$, in the same way that one constructs them for e^x. Here are the Taylor series for $\cos x$ and $\sin x$:

$$\cos x = 1 - \frac{x^2}{2!} + \frac{x^4}{4!} - \frac{x^6}{6!} + \frac{x^8}{8!} - \frac{x^{10}}{10!} + \cdots$$

and

$$\sin x = \frac{x}{1!} - \frac{x^3}{3!} + \frac{x^5}{5!} - \frac{x^7}{7!} + \frac{x^9}{9!} - \cdots .$$

For example, if you take the first four terms of the series for $\cos x$ for $x = \pi/4$, you get 0.707103, which is quite a good approximation to $\cos(\pi/4) = \sqrt{2}/2 \approx 0.707107$.

So now, using the Taylor series for e^x, we have a way of calculating e^x for any real x. But what should e^z be if z is complex? Keep in mind that real numbers are also complex numbers—they just have an imaginary part of 0, i.e., they are $a + bi$ with $b = 0$. So whatever definition we give to e^z for complex z's must agree with the definition of e^x for real x if z is real. Since we've already defined e^3, e^{3+0i} must equal e^3. Given that we have a Taylor series for e^x when x is real, how would *you* define e^z if z is complex? Personally I would just take the Taylor series we've already given and use it to define e^z, that is, I would *define* e^z to be

$$e^z = 1 + \frac{z}{1!} + \frac{z^2}{2!} + \frac{z^3}{3!} + \frac{z^4}{4!} + \frac{z^5}{5!} + \frac{z^6}{6!} + \cdots .$$

This is indeed how e^z is defined for complex z!

Now that we've built up the machinery, we're ready to prove Theorem 5.4. It's not hard.

Proof of Theorem 5.4 We must prove that for any real θ, $e^{\theta i} = \cos\theta + i \cdot \sin\theta$. From the Taylor series for e^z, we have

$$e^z = 1 + \frac{z}{1!} + \frac{z^2}{2!} + \frac{z^3}{3!} + \frac{z^4}{4!} + \frac{z^5}{5!} + \frac{z^6}{6!} + \cdots ,$$

and so

$$e^{\theta i} = 1 + \frac{(\theta i)}{1!} + \frac{(\theta i)^2}{2!} + \frac{(\theta i)^3}{3!} + \frac{(\theta i)^4}{4!} + \frac{(\theta i)^5}{5!} + \frac{(\theta i)^6}{6!} + \frac{(\theta i)^7}{7!} + \cdots$$

$$= 1 + \frac{i\theta}{1!} + \frac{i^2\theta^2}{2!} + \frac{i^3\theta^3}{3!} + \frac{i^4\theta^4}{4!} + \frac{i^5\theta^5}{5!} + \frac{i^6\theta^6}{6!} + \frac{i^7\theta^7}{7!} + \cdots .$$

Now we use the fact that $i^2 = -1$, $i^3 = -i$, $i^4 = 1$. Since $i^4 = 1$, the numbers cycle: $i^5 = i^1 = i$, $i^6 = i^2 = -1$, $i^7 = i^3 = -i$, $i^8 = i^4 = 1$, etc. Therefore:

$$e^{\theta i} = 1 + \frac{i\theta}{1!} - \frac{\theta^2}{2!} - \frac{i\theta^3}{3!} + \frac{\theta^4}{4!} + \frac{i\theta^5}{5!} - \frac{\theta^6}{6!} - \frac{i\theta^7}{7!} + \cdots .$$

Now separate out the terms involving i from those not involving i:

$$e^{\theta i} = \left[1 - \frac{\theta^2}{2!} + \frac{\theta^4}{4!} - \frac{\theta^6}{6!} + \cdots \right] + i \left[\frac{\theta^1}{1!} - \frac{\theta^3}{3!} + \frac{\theta^5}{5!} - \frac{\theta^7}{7!} + \cdots \right].$$

Comparing this to the Taylor series for $\cos\theta$ and $\sin\theta$, we see that the first bracketed expression is $\cos\theta$ while the second one is $\sin\theta$, and we are done. ∎

5.5   Puzzle

There are two vertical towers separated on the ground by c meters. The first tower T_1 is a meters high, the second, T_2, b meters high. On the Euclidean plane, you can think of T_1 as running between $(0, 0)$ and $(0, a)$ and T_2 as running between $(c, 0)$

and (c, b). You want to choose an x, $0 < x < c$, such that a chain run from the top of T_1 to the point on the ground x meters from the bottom of T_1 (and $c - x$ meters from the bottom of T_2) –this is the point $(x, 0)$– and back up to the top of T_2 has minimum length. By the Pythagorean Theorem, the length of the chain will be $\sqrt{a^2 + x^2} + \sqrt{(c - x)^2 + b^2}$. What should x be?

5.6   Exercises

Exercise 5.1 Let $z_1 = a + bi$ and $z_2 = c + di$ be two complex numbers. Show that $|z_2 + z_2| \leq |z_1| + |z_2|$. This actually has nothing to do with i. It says that the length of the line segment from $(0, 0)$ to $(a + c, b + d)$ is at most the length of the line segment from $(0, 0)$ to (a, b) plus the length of the line segment from (a, b) to $(a + c, b + d)$. This is called the *triangle inequality* since it says that the length of any one side of a triangle is at most the sum of the lengths of the other two sides.

Exercise 5.2 Find all complex roots $z = a + bi$ of $z^2 = i$.

Exercise 5.3 * Find all three complex roots $z = a + bi$ of $z^3 = 1$.

Exercise 5.4 Find all four complex roots $z = a + bi$ of $z^4 = 1$.

Exercise 5.5 What is the value of e^i? Give the real and imaginary parts to six decimal places.

Exercise 5.6 Over the reals, $\ln(x)$ takes any positive x and returns the unique real y such that $e^y = x$. Find infinitely many *complex* numbers z such that $e^z = 1$, showing that $\ln(1)$ can have infinitely many values.

Exercise 5.7

(a) Write a Python program that estimates $\sin x$ by computing the first d terms of the Taylor series.
(b) We know that $\sin(2\pi) = 0$. How many terms of the Taylor series are needed to get an estimate of $\sin(2\pi)$ which has absolute value at most 0.01?

Exercise 5.8 You might have asked how one knows that the Taylor series for e^x converges (i.e., approaches one value as you take more and more terms). After all, the series $1 - 1 + 1 - 1 + 1 - 1 + \cdots$ doesn't converge: if you take an odd number of terms, the sum is 1; if you take an even number of terms, the sum is 0.

It is known that a series will converge if the absolute value of the ratio between consecutive terms approaches a value r which is *strictly less* than 1. Notice that this (fortunately) fails for the annoying series $1 - 1 + 1 - 1 + 1 - 1 + \cdots$: the absolute value of each term divided by the previous one is exactly 1, and so approaches a limit of 1, which is not less than 1.

(a) In the Taylor series for e^x, the nth term, for $n = 0, 1, 2, 3, \ldots$, is $x^n/(n!)$. Show that for any x, the absolute value of the ratio between consecutive terms approaches 0, which means that the series converges.

(b) Show that the absolute ratios approach 0 in the Taylor series for $\cos x$.

(c) Do the same for $\sin x$.

Exercise 5.9 Now that we have the Taylor series for e^x, you and I are going to prove together that e is irrational. Recall that

$$e = e^1 = 1 + \frac{1}{1!} + \frac{1}{2!} + \frac{1}{3!} + \cdots .$$

Suppose for a contradiction that $e = a/b$ with a, b integers with $b \geq 2$. Write $e = p + s$ with p for "prefix" and s for "suffix" as follows:

$$p = 1 + \frac{1}{1!} + \frac{1}{2!} + \frac{1}{3!} + \cdots + \frac{1}{b!}$$

and

$$s = \frac{1}{(b+1)!} + \frac{1}{(b+2)!} + \frac{1}{(b+3)!} + \cdots .$$

Let $\alpha = b!(e - p) > 0$.

(a) Show that α is an integer.

Now

$$\alpha = b!(e - p)$$

$$= b!s$$

$$= \frac{b!}{(b+1)!} + \frac{b!}{(b+2)!} + \frac{b!}{(b+3)!} + \cdots$$

$$= \frac{1}{b+1} + \frac{1}{(b+1)(b+2)} + \frac{1}{(b+1)(b+2)(b+3)} + \cdots$$

$$< \frac{1}{b+1} + \frac{1}{(b+1)^2} + \frac{1}{(b+1)^3} + \cdots .$$

(b) This is an infinite geometric series. Evaluate it exactly and show that it's at most 1/2.

(c) Find a contradiction. The existence of the contradiction implies that the assumption we made, that e is rational, must be false. ∎

5.6.1 Hint

5.3 The obvious one is $z = 1$. This means that $z^3 - 1$ can be written as $z - 1$ times a quadratic polynomial in z. (This is not obvious.) How can one use this fact to find the other two roots?

Infinities and Diagonalization

6

6.1 Finite Sets

How many positive integers are there? I know you know there are "infinitely many." They "go on forever." Can you say that the number is "infinity?" Is "infinity" a number? If so, is there really only *one* infinity?

We will discuss these questions in this chapter. You may be surprised to learn that there are many different infinities. How could it make sense to talk about different infinities? Isn't infinity just infinity?

Suppose I have a bunch of oranges and you have a bunch of apples. How can we find out who has more, without counting the apples and oranges? We don't want to count them because later we will use the same idea on infinite sets, which cannot be counted.

Here's what we do. Simultaneously, you take one apple from your set and place it on a table, and I take one orange from my set and place it on the table. Then you take another apple from your set and place it on the table while I take another orange from my set and place it on the table. Then we do the same thing again, and again. If your bunch of apples lasted longer than my bunch of oranges, then you had more; if my oranges lasted longer, than I had more; and if we ran out at the same time, then we had collections of fruit of the same size. Notice that we don't need to have *counted* the sets of fruit, or even to have a concept of *number*, in order to do this. We can determine which set is larger, or that they're the same size, without knowing the size of either set.

© The Author(s), under exclusive license to Springer Nature Switzerland AG 2023   71
H. Karloff, *Mathematical Thinking*, Compact Textbooks in Mathematics,
https://doi.org/10.1007/978-3-031-33203-6_6

A *mapping* or *function* from a set A to a set B is a procedure that takes an element of set A and assigns to it an element of set B. You've probably seen this in school. It's important to realize that you can't switch the A and the B. For example, if $\mathbb{N}$ is the set $\{0, 1, 2, 3, 4, \ldots\}$ of nonnegative integers and $\mathbb{E}$ is the set $\{0, 2, 4, 6, 8, \ldots\}$ of even nonnegative integers, then here are three examples of mappings from $\mathbb{N}$ to $\mathbb{E}$:

1. mapping f which takes a nonnegative integer x and assigns it $f(x) = 2x$ (e.g., $f(5) = 10$),
2. mapping g which takes a nonnegative integer x and assigns it $g(x) = 4x$ (e.g., $g(5) = 20$), and
3. mapping h which takes a nonnegative integer x and assigns it $h(x) = 6$.

All of these are mappings from $\mathbb{N}$ to $\mathbb{E}$, since the "output" $f(x)$, $g(x)$, or $h(x)$, is always a nonnegative even integer.

Mappings f and g have the nice property that different "inputs" x give different "outputs" $f(x)$ or $g(x)$, respectively. Functions like these two are called *one-to-one*. By contrast, h is clearly not one-to-one: all the inputs x give the same $h(x)$, namely, 6.

The first function f has one more nice property: it "hits" all the nonnegative even integers, that is, $f(0) = 0$, $f(1) = 2$, $f(2) = 4$, $f(3) = 6, \ldots$ and in this way it hits all the nonnegative even integers. By contrast, g does not: $g(0) = 0$, $g(1) = 4$, $g(2) = 8, \ldots$ but there's no g that hits 2. Clearly h does not, either. A function f from A to B that "hits" all the elements of B is called *onto*. By the way, onto functions are allowed to "hit" elements of B more than once. The only requirement is that they "hit" all elements of B at least once.

For finite sets, we know what *size* of a set is, it's just the number of elements in the set. It's not so simple with infinite sets.

Let's think about how we found out whether you had more apples or I had more oranges (or neither). We paired up the apples and oranges for a while, until we ran out. If there were 0 or more oranges left over at the end, then there were at least as many oranges as apples. Similarly, if there were 0 or more apples left over at the end, then there were at least as many apples as oranges.

Let's try to do something similar with infinite sets. We can't talk about what happens "at the end" because there will be no end. Still, let's see if we can come up with a reasonable definition for what it would mean for there to be "at least as many oranges as apples" even when the sets of apples and oranges are infinite.

Let's start pairing up apples and oranges. If we pair *all* the apples with oranges, different apples getting paired with different oranges (of course), and there are 0 or more oranges left over, then we have paired up every apple with an orange (but possibly not every orange with an apple), so there is a one-to-one mapping from the whole set of apples to the whole set of oranges. (You don't get to use the same orange twice.) In this case, the number of oranges is at least as large as the number of apples.

Similarly, if there is a one-to-one mapping from the set of oranges to the set of apples, then the number of apples is at least as large as the number of oranges.

What have we learned? We've learned that in the case of two finite sets A and B, B is at least as large as A if, and only if, there is a one-to-one mapping from A to B.

By the way, when are the two sets A and B of the same size? They are of the same size if, and only if, there is a mapping from A to B which is one-to-one *and also onto*. A mapping from A to B which is one-to-one and onto is called a *bijection*. For example, the mapping f above from the set $\mathbb{N}$ of nonnegative integers to the set $\mathbb{E}$ of nonnegative even integers is a bijection. Mappings g and h are not onto and hence not bijections.

By the way, we will need these two facts, which are easy to prove and which I will leave as exercises.

Fact. Let A, B be sets. If there is a bijection from A to B, then there is a bijection from B to A. Hence we say there is a bijection *between A and B*.

Proving this is Exercise 6.1.

Fact. Let A, B, C be sets. If there are bijections from A to B and from B to C, then there is a bijection from A to C.

Proving this is Exercise 6.2.

6.2   Infinite Sets

For sets which may be infinite, we will adopt the same ideas. Given two sets A and B, first we will define what it means for set A to be "no larger than B." Notice that I am not saying "smaller"; I am saying "no larger." Then we will define what it means for A and B to be "the same size."

By the way, there is an interesting point here. For a finite set A, we have the concept of "size" of the set, which is a nonnegative integer. The size of a finite set A depends only on that set A. To see whether finite set A is no larger than finite set B, we can just ask if the size of A is no larger than the size of B. Notice that we can find out which of two finite sets A and B is larger by *separately* computing first the size of A from A and the size of B from B, and then comparing the sizes. In the case of infinite sets, we will not have a concept of size. *We will not define size of an infinite set.*

Now, what shall it mean to say that A is no larger than B? We'll use the pairing trick we illustrated above.

Definition 6.1 We say set A is *no larger than* B if there is a one-to-one mapping from A to B.

Here are some examples:

1. The set $\mathbb{Z}$ of integers is no larger than the set $\mathbb{R}$ of reals. Since every integer is also a real, this one is easy to prove. Just set $f(x) = x$ for every integer x.

2. This one is more interesting. Let A be the set of even nonnegative integers and let B be the set of nonnegative integers which are multiples of 4. That is, $A = \{0, 2, 4, 6, 8, \ldots\}$ and $B = \{0, 4, 8, 12, 16, \ldots\}$. Clearly B is a (proper) subset of A, so you might be surprised to hear that A is no larger than B. Indeed, define the map $f(x) = 2x$ for $x \in A$. This map f is clearly one-to-one (and onto), so A is no larger than B.

3. Here's a really interesting (non-)example, which is a twist on the first example. Let A be the set $\mathbb{R}$ of reals and let B be the set $\mathbb{Z}$ of integers. Clearly $B = \mathbb{Z}$ is a (proper) subset of $A = \mathbb{R}$. Is A no larger than B, as in the previous example? We will see later that it is *not*: in fact, $\mathbb{R}$ is larger than $\mathbb{Z}$.

Now that we know what it means for A to be no larger than B, let's define what it means for A to be "the same size" as B and for A to be "smaller" than B. In the case of finite sets, we saw that two sets were the same size if, and only if, there was a bijection (a one-to-one and onto mapping) between them. Let's use the same definition for general sets.

Definition 6.2 Sets A and B are *the same size* if there is a bijection between them.

I am defining what it means for two sets to be of the same size *without defining the size of an individual set*. Here's an interesting example. Take $A = \{0, 2, 4, 6, 8, \ldots\}$ and $B = \{0, 4, 8, 12, 16, \ldots\}$. The mapping $f(x) = 2x$ from A to B is a bijection, so, perhaps surprisingly, A and B have the same size. This is something you probably wouldn't have guessed, since B is a proper subset of A.

The set $\mathbb{Z}$ of integers and the set $\mathbb{R}$ of reals do *not* have the same size. We will see a beautiful proof later of this interesting fact, which is far from obvious. How would you prove the *nonexistence* of a bijection between $\mathbb{Z}$ and $\mathbb{R}$?

Here's a useful definition. Let $\mathbb{P} = \{1, 2, 3, 4, \ldots\}$ be the set of positive integers.

Definition 6.3 A set A is *countable* if it has the same size as the set $\mathbb{P}$ of positive integers. In other words, a set A is *countable* if there is a bijection from $\mathbb{P}$ to A.

For example:

Theorem 6.1 *The set A of all even positive integers is countable.*

Proof $A = \{2, 4, 6, 8, \ldots\}$. We need to find a bijection from $\mathbb{P} = \{1, 2, 3, 4, \ldots\}$ to A. The bijection f from $\mathbb{P}$ to A is $f(x) = 2x$. This mapping f is clearly one-to-one and onto, and hence a bijection, proving that A is countable. ∎

An even more interesting (non-)example is that the set $\mathbb{R}$ of reals is *not* countable.

Definition 6.4 Set A is *smaller than* set B if A is no larger than B, and it is not true that A and B have the same size.

Here is a really interesting question. Suppose that A is no larger than B and that B is no larger than A. In the case of finite sets, one could conclude that A and B have the same size. Is the same true for general sets? In fact, it *is* true, but it is far from obvious and surprisingly hard to prove, hard enough that I will not include a proof here. You can find a proof of the Schröder-Bernstein theorem on Wikipedia.

Theorem 6.2 *The Schröder-Bernstein or Cantor-Schröder-Bernstein Theorem. If A, B are two sets such that there are a one-to-one mapping from A to B and a one-to-one mapping from B to A, then there is a bijection between A and B.*

As I said, this is not obvious at all.

6.3   The Set of Rationals Is Countable

Remember from Chap. 4 that the *rational numbers* are just the "ratios" (hence the word "rational") or fractions a/b of integers a, b in which $b > 0$. Normally b is allowed to be negative, but there is no harm in requiring b to be positive, since we can negate both the denominator and numerator if b is negative. The positive rational numbers can be all written as a/b where both a and b are positive integers. Let us use $\mathbb{Q}$ to denote the set of all rationals and $\mathbb{Q}^+$ to denote the set of positive rationals.

(I remember when I first saw the phrase "rational number" in ninth grade I thought those were numbers that "made sense" or were "logical," since those are meanings of "rational" in standard English. I was quickly corrected.)

A positive rational x has many different representations as a/b with both a and b positive. For example, $2/3 = 4/6 = 6/9 = 8/12 = \cdots$; they are all the same rational. Still, there appear to be many more positive rationals than positive integers. You use *two* integers to describe a rational. It may come as a surprise, then, that the positive rationals are no more numerous than the positive integers! The set of positive rationals and the set of positive integers have the same size. Specifically, we will prove.

Theorem 6.3 *There is a bijection between the set $\mathbb{P}$ of positive integers and the set $\mathbb{Q}^+$ of positive rationals. Therefore $\mathbb{Q}^+$ is countable.*

Here we go.

Proof Write down a two-dimensional array M in which the rows are labeled $1,2,3,4,\ldots$ in order and the columns are labeled $1,2,3,4,\ldots$ in order as well. As in the small part below of the table, in the cell in the row labeled a and the column labeled b, write down the fraction a/b, if a/b is in lowest terms, and, if not, leave the cell blank. For example, in row 2 and column 3, write 2/3. Write nothing in row 2 and column 4, because 2/4 is not in lowest terms. Since every positive rational number x has exactly one representation as $x = a/b$ with a and b both positive and

a/b in lowest terms, every positive rational number appears exactly once in the table (and no irrationals appear, of course).

Here is a small part of the table and bijection f:

```
         1               2               3             4             5           6
1    1/1 = f(1)      1/2 = f(3)      1/3 = f(5)    1/4 = f(9)    1/5 = f(11)  1/6 = f(17)
2    2/1 = f(2)                      2/3 = f(8)                  2/5 = f(16)
3    3/1 = f(4)      3/2 = f(7)                    3/4 = f(15)      3/5
4    4/1 = f(6)                      4/3 = f(14)                     4/5
5    5/1 = f(10)     5/2 = f(13)        5/3          5/4                         5/6
6    6/1 = f(12)                                                    6/5
```

We have to give a bijection f, that is, a one-to-one and onto mapping, from $\mathbb{P}$ to $\mathbb{Q}^+$. Since $\mathbb{P} = \{1, 2, 3, 4, \ldots\}$, I will write down $f(1), f(2), f(3), f(4), \ldots$, in that order. It is important to understand at this point that we have to prove that such an f exists; we don't have to give a formula for it.

The way to get the bijection is to start at the upper-left corner of M and then start traversing the lower left-to-upper right "diagonals" afterward, one at a time, skipping blank cells. Start in the upper-left corner and write down its entry: $1/1$, that is, $f(1) = 1/1$. Then go down one cell to $2/1$, that is, $f(2) = 2/1$. Then continue diagonally upward and to the right (in the "northeast" direction) until hitting the top. That is, after $2/1$, go to $1/2 = f(3)$. Then jump to the third entry in the first column, $3/1 = f(4)$, and then move northeast, skipping the 2nd entry in the 2nd row, and then visiting $1/3 = f(5)$. Next, jump down to $4/1 = f(6)$, and then continue northeast to $3/2 = f(7)$, $2/3 = f(8)$, and $1/4 = f(9)$. Next, jump down to $5/1 = f(10)$, skip $4/2$, $3/3$, and $2/4$, and then go to $1/5 = f(11)$. Continue in this way forever.

So our mapping f begins with $< f(1), f(2), f(3), f(4), f(5), f(6), f(7),$ $f(8), f(9), f(10), f(11), \ldots . > = < 1/1, 2/1, 1/2, 3/1, 1/3, 4/1, 3/2, 2/3, 1/4,$ $5/1, 1/5, \ldots >$.

It's clear that the mapping f exists. After all, if you give me a positive integer n, like $n = 232491302$, I can tell you what $f(n)$ is (though it might take me a while). The mapping f is one-to-one because we've only included fractions in lowest terms, and it's onto because we've included all positive rationals. The proof is complete.   ∎

The positive rational numbers can be viewed as a subset of the pairs (a, b) of positive integers. A similar proof to this last proof shows that the set of all triples (a, b, c) of positive integers is also countable, as is the set of all quadruples (a, b, c, d) of positive integers, quintuples, etc.

6.4   The Set of Reals Is Uncountable

What about the reals? We just claimed that the sets of pairs, or triples, or quadruples, etc., of positive integers are countable. There's no obvious way to represent an arbitrary real as a sequence of two, or three, or four, etc., positive integers. However, each real between 0 and 1 (including 0 but not 1) has a decimal representation starting with a decimal point followed by an *infinite* sequence of digits $0, 1, 2, 3, \ldots, 9$. In fact, if you know about binary representations, though you don't have to, each such real can be represented as an infinite sequence of 0's and 1's. Rationals use pairs of integers. Reals use fewer symbols –ten in the decimal case and two in the binary case– but they need infinitely many symbols instead of the two needed, one for the numerator and one for the denominator, in the case of rationals. So what's the story here? Are the reals countable or not? It turns out that they are not. Let's work specifically with the set S of reals which are greater than or equal to 0 but less than 1. The mathematical notation for this set is $[0, 1)$, where the square bracket next to the 0 means that 0 is in the set and the parenthesis next to the 1 means that 1 is not in the set.

Theorem 6.4 *The set* $S = [0, 1)$ *of reals* x *such that* $0 \le x < 1$ *is not countable.*

You may be asking yourself, "How could one prove that a set is *not* countable?" After all, to prove a set S *is* countable, one just has to present a mapping f from the positive integers $\mathbb{P}$ to S and then prove that f is one-to-one and onto. How would one prove that a set is *not* countable? It would not suffice to show that a particular mapping f doesn't work, since a different mapping g might work. One has to rule out all possible mappings at once.

The trick is to use a proof by contradiction. We will start off by pretending that a one-to-one and onto mapping f exists from $\mathbb{P}$ to S, and then show that in fact f is not onto, in other words, that f misses at least one number in S. Since the proof will work for any mapping f, the conclusion is that the desired mapping cannot exist.

Let's do it.

Proof Suppose, for a contradiction, that f is a one-to-one and onto mapping from the set $\mathbb{P}$ of positive integers to the set S of reals x that are at least 0 and less than 1. Each real $x \in S$ has a decimal expansion. Choose a decimal expansion for each such x. Remember, from Chap. 4, that each real has exactly one decimal expansion, unless it has a decimal expansion ending in all 0's or all 9's, in which case it has a second decimal expansion ending in all 9's if the first expansion ended in all 0's and a second decimal expansion ending in all 0's if the first expansion ended in all 9's, but no third decimal expansion. In this case, use the expansion ending in all 0's. Build a table whose first row is the decimal expansion of the real $f(1)$, whose second row is the decimal expansion of the real $f(2)$, whose third row is the decimal expansion of $f(3)$, etc. For example, if $f(1) = 0.53$, $f(2) = 2/3$, $f(3) = \pi/10$, and $f(4) = (\sqrt{2})/10$ where $\sqrt{2} \approx 1.414$, then the table would begin:

0.530000000000000...
0.666666666666666...
0.314159265358979...
0.141421356237309...

There is one row for every positive integer.

Since the mapping is one-to-one and onto, each real number in S should appear *exactly once* in this table. I am going to convince you now that, in fact, the table is missing at least one real number, and hence the mapping is not onto, a contradiction.

Which number is missing? Look down the "diagonal" of the table. That is, look at the *first* digit after the decimal point in the *first* number, the "5"; the *second* digit after the decimal point in the *second* number, the "6"; the *third* digit after the decimal point in the *third number*, the "4"; the *fourth* digit after the decimal point in the *fourth* number, the "4"; etc. Write down the infinite sequence of these digits as a real number: $0.5644\ldots.$ Call this real number d, "d" for "diagonal." Because of the way d was constructed, for any positive integer i, the ith digit of d is the ith digit of $f(i)$. For the table drawn above, d would start $0.5644.$

Now –and this is the main idea– change every digit to a different digit, but not to a 0 or a 9. There are ten digits $0,1,2,\ldots,9$, and one of them is already used, leaving nine others, and I just told you to avoid 0 and 9. That leaves at least seven possible new digits. You can choose any digit you wish of the seven remaining. Call the new decimal number you got d'. The numbers d and d' differ in every decimal digit. For the example above, d' might begin $0.1836\ldots.$ The key point is that *all* the digits of d and d' differ.

Because we were careful to avoid 0's and 9's, clearly d' does not end in an infinite string of 0's or an infinite strings of 9's, and it follows that d' has only one decimal representation.

Now we get to the meat of the argument. The mapping f is supposed to be onto. This means that d' appears on some row, say, row i', of the table, that is, $f(i') = d'$. However, in the table we constructed, the i'th digit in the i'th row equals the i'th digit of d. Since *all* the digits of d and d' differ, $f(i')$ *has the wrong digit in the i'th place to be d'*! This means that in fact $f(i') \neq d'$! This is the contradiction.   ∎

Let me continue the example here, since it might help. In the example above, d began 0.5644 and d' began 0.1836.

d' can't be $f(1)$, because $f(1)$ has a 5 in its first position, and d' has a 1 in its first position.

d' can't be $f(2)$, because $f(2)$ has a 6 in its second position, and d' has an 8 in its second position.

d' can't be $f(3)$, because $f(3)$ has a 4 in its third position, and d' has a 3 in its third position.

d' can't be $f(4)$, because $f(4)$ has a 4 in its fourth position, and d' has a 6 in its fourth position.

Continuing in this way, we see that d' can't be $f(i')$ for any i'.

Because the set S is not countable, we say it's *uncountable*. By the way, if the set $\mathbb{R}$ were countable, S would be so as well, so it follows that the full set $\mathbb{R}$ of reals is also uncountable. This is a very interesting fact. While the number of positive integers equals the number of rationals, the number of reals is strictly greater.

This proof was originally published in 1891 by Georg Cantor. Because of the use of the diagonal, it's known as *Cantor's diagonalization argument*. In the next section, we will see how a diagonalization argument can be used to prove a fundamental theorem in computer science, specifically, that it is impossible to determine whether an arbitrary computer program will ever halt on a given input.

6.5   The Halting Problem Is Undecidable

When you run a computer program, there are only two things that can happen: it stops after a finite length of time –this is known as *halting*– or it runs forever. It can stop for multiple reasons, perhaps because it worked and produced the answer you wanted, perhaps because of an error in the program or in the data. The case of running forever can be frustrating, as you don't know if it would really run forever or if it's just about to halt. It would surely be useful to have a program, say, h for "halting tester," to which you feed a second program A and an input x to that program, to which h responds with 1 if A would eventually halt when run on input x and with 0 if A would run forever on input x. Program h would have to halt on all inputs, since a program that itself runs forever wouldn't be very helpful.[1] This rules out the naive solution h of just running program A on input x, since if A ran forever on x, then so would h.

In this section, we will prove the celebrated result of Alan Turing[2] that no such program exists. Even if the program is given an arbitrary amount of time and space, it is impossible to solve this problem. It is *undecidable.* This theorem is one of the most fundamental in theoretical computer science. Its proof uses Cantor's diagonalization technique.

Theorem 6.5 *The halting problem is undecidable. That is, there is no computer program h that takes a program program_string and an input input_string, always halts, and outputs 1 if program program_string halts on input input_string, and 0 otherwise.*

Proof Suppose, for a contradiction, that there were such a program h. Let's suppose it's written in Python. Write a new Python program $f.py$, which uses h:

```
def h(program_string, input_string):
    # Program h, which always halts, goes here.  It outputs
    # 1 if program "program_string" halts when run on input "input_string",
    # and 0 otherwise.

# The main program starts here.
# Enter a string at the keyboard.
p = input()
if h(p, p) == 1: # If h says p will halt on input p, then run forever.
  while True:
    pass
# Otherwise, just halt.
```

---

[1] This is not exactly true. A program h run every day which ran forever in practice only once a decade would still be helpful.

[2] (Portrayed by Benedict Cumberbatch in the 2014 movie, "The Imitation Game").

Any text after a "#" on a line is a comment and is ignored. First we include the code for the function h, which we've assumed exists. Then the main program begins with a request for the user to enter a string p at the keyboard. Next, the program calls function h, strangely with both *program_string* and *input_string* equaling p. This may seem odd, but it's akin to Cantor's diagonalization argument, in which for the ith integer he changed the ith digit. Now we have to "change the ith digit": if $h(p, p) = 1$, that is, program h says that program p halts on input string p, then we put the program into an infinite loop ("while True:" means "run forever," and "pass" means "do nothing"). If $h(p, p) = 0$, which is the only other possibility, then we skip the infinite loop and halt.

Now we derive a contradiction by running the program on itself. More precisely, we start up the program $f.py$ and, sitting at the keyboard, we enter the full program $f.py$ as one string p (omitting the comments). That string starts with

```
def h(program_string, input_string):\n
```

and ends with

```
if h(p, p) == 1:\n   while True:\n      pass\n
```

(The backslash-n's are newlines.) There are two cases:

1. If $h(f.py, f.py) = 1$, then program $f.py$ enters an infinite loop so $f.py$ runs forever when run on input string $f.py$. But $h(f.py, f.py) = 1$ says that program $f.py$ halts when run on input string $f.py$, which is wrong. It follows that we must have $h(f.py, f.py) = 0$.
2. However, if $h(f.py, f.py) = 0$, then program $f.py$ halts when run on input string $f.py$. But $h(f.py, f.py) = 0$ says that program $f.py$ runs forever on input $f.py$, a contradiction.

The only assumption we made, which h exists, must therefore be false.   ∎

6.6   Puzzle

Take a 10×10 square and divide it up into 100 1×1 cells in the obvious way. Remove the lower-left and upper-right cells, leaving 98 cells. You have a supply of 49 2×1 dominoes, each one of which can cover two horizontally or vertically adjacent cells. Can you cover the 98 remaining cells with the 49 dominoes?

6.7   Exercises

Exercise 6.1 Show that if A and B are sets and there is a bijection from A to B, then there is a bijection from B to A.

Exercise 6.2 Let A, B, C be sets. Show that if there is a bijection from A to B and one from B to C, then there is a bijection from A to C.

Exercise 6.3 * Let $\mathbb{P} = \{1, 2, 3, 4, \ldots\}$ be the set of positive integers. Give a bijection between the set $\mathbb{P}$ and the set of pairs of positive integers, thus proving that the set of pairs of positive integers is countable.

Exercise 6.4 Show that the set F of *finite* sequences (of any length) of positive integers is countable. That is, F consists of all sequences of length 0 of positive integers, all sequences of length 1 of positive integers, all sequences of length 2 of positive integers, etc.

Exercise 6.5 * We know that there are irrational numbers such as π and $\sqrt{2}$. The set of real numbers can be divided into the *algebraic* numbers, like $4/3$ or $\sqrt{2}$, and the *nonalgebraic* or *transcendental* numbers like π. A real number α is *algebraic* if there is a nonzero polynomial $p(x)$ with integer coefficients such that $p(\alpha) = 0$. For example, if $\alpha = a/b$ with a, b integers and $b \neq 0$, then for $p(x) = bx - a$, $p(\alpha) = 0$, and so the rational a/b (like every rational) is algebraic. But some irrationals, like $\sqrt{2}$, are also algebraic. For $\alpha = \sqrt{2}$ and $p(x) = x^2 - 2$, we clearly have $p(\alpha) = 0$ and so $\sqrt{2}$ is algebraic. By contrast, it so happens that there is no nonzero polynomial $p(x)$ with integer coefficients such that $p(\pi) = 0$ –this is not obvious– and so π is transcendental.

Show that the set of algebraic numbers is countable.

Exercise 6.6 * The *power set* of a set S is the set of all of S's subsets.

Write a Python program that takes a positive integer n as input and outputs the power set of $\{0, 1, 2, 3, \ldots, n - 1\}$. For example, for $n = 3$, your program should output something like

```
{}
{0}
{1}
{2}
{0 1}
{0 2}
{1 2}
{0 1 2}
```

The elements of each set can be permuted and the rows themselves can be permuted, too. Output the sets and the elements in each set in whatever order is most convenient for you.

Exercise 6.7 * An *open interval* (a, b) is the set of all reals greater than a and less than b, where $a < b$. For any $\epsilon > 0$, find, for each rational q strictly between 0 and

1, an open interval of reals centered at q so that the sum (over those rationals) of the lengths of all the intervals is no more than ϵ.

Since the length of the interval $[0, 1]$ between 0 and 1 is 1, this informally shows that the fraction of reals between 0 and 1 which are rationals is at most ϵ, for all $\epsilon > 0$.

Exercise 6.8 Write a Python program that takes n and produces a list of the first n positive rationals, in the order given by the proof of Theorem 6.3 that the set of positive rationals is countable.

6.7.1 Hints

6.3 Use the idea of the proof that the set of rationals is countable.

6.5 Use the result of Exercise 6.4 and the fact that a nonzero polynomial of degree d has at most d distinct roots.

6.6 Use the fact that each of the 2^n subsets of $\{0, 1, 2, \ldots, n-1\}$ can be represented by a 0–1 string of length n.

6.7 Use the fact that the set of rationals is countable.

Binary Search

7

7.1   Binary Search

Have you ever played this game? Your friend has an integer between, say, 1 and 15, and asks you to guess it. If your guess is correct, your friend shouts "Got it!", and the game stops. If not, your friend tells you whether the hidden number is larger or smaller than your guess, and you guess again.

For example, suppose your friend's secret number is 2. Without knowing that, of course, you guess 10. Your friend says, "Lower." You guess 8. Your friend again says "Lower." You guess 1. "Higher" this time. You guess 5, which elicits "Lower." Then you guess 3, which elicits "Lower." Next you guess 2, your friend responds "Got it!", and the game is over. It took you 6 guesses (10, 8, 1, 5, 3, 2).

You keep making queries until you finally get "Got it!". Even if you know what the missing number is, you have to use one query to force your friend to respond, "Got it!".

How can you play the game better? In fact, how can you play it optimally? *Optimally* means *in the best possible way*.

A (closed) *interval* is the set of all reals between two given reals a and b, with $a \leq b$, including both endpoints. We will denote the interval from a to b, including both endpoints, as $[a, b]$. For example, $[9, 14.3]$ is the set of all reals between 9 and 14.3, including both 9 and 14.3. The key point about this game is that you can always represent the location of the hidden number as lying in an interval. At the beginning, the interval is $[1, 15]$. Suppose, as above, your first guess is 10, to which the answer is "Lower." Then you know that the hidden number is in $[1, 9]$. After your next guess of 8, and the answer of "Lower," you know the hidden number is in $[1, 7]$.

It is important to remember that the hidden number is known to be in the interval and that, on the other hand, as far as you know, it could be *any* of the integers in the interval.

© The Author(s), under exclusive license to Springer Nature Switzerland AG 2023
H. Karloff, *Mathematical Thinking*, Compact Textbooks in Mathematics,
https://doi.org/10.1007/978-3-031-33203-6_7

In the play of the game above, six guesses sufficed. How can one play the game so as to guess the hidden number in four (or fewer) guesses? Here's the trick: always guess the *middle* number in the interval (or as close to the middle as possible, if there's no middle integer).

So how should one play the game? Initially the interval is [1, 15], so the first guess is 8. The answer is "Lower." The interval becomes [1, 7]. The middle number, and hence the guess, is 4, to which the answer is "Lower." The interval becomes [1, 3], so that the next guess is 2, the response is "Got it!", and the game is over. We needed only three guesses.

Think for a moment about what would have happened if the hidden number had been 3. The game would have been played exactly as before, except when the guess was 2, the answer would have been "Higher," and the hidden interval would have become [3, 3]. The next guess would have been 3 and the game would have been over, with four guesses needed.

Let's see why four guesses suffice. Initially the interval is [1, 15], which contains 15 integers. The first guess is 8. Either 8 is right and we're done, or it's not. If it's not, the new interval becomes either [1, 7] (if the answer was "Lower") or [9, 15] if the answer was "Higher." Either way, the new interval contains seven integers.

The second guess will lie right in the middle, either of [1, 7] (giving a guess of 4) or of [9, 15] (giving a guess of 12). The key point is that, unless we're lucky and have already gotten the hidden number, the new interval will have three integers (it will be either [1, 3], [5, 7], [9, 11], or [13, 15]).

The third guess, again in the middle, will either be lucky or shrink the interval to one integer (either [1, 1], [3, 3], [5, 5], [7, 7], [9, 9], [11, 11], [13, 13], or [15, 15]). The next guess will definitely succeed. Four guesses will have sufficed.

What would have happened if the hidden number were known to be in [1, 31] instead of [1, 15]? (And why was I asking about [1, 15] before and [1, 31] now? Is it coincidence that $15+1 = 16 = 2^4 = 2 \cdot 2 \cdot 2 \cdot 2$ and $31+1 = 32 = 2^5 = 2 \cdot 2 \cdot 2 \cdot 2 \cdot 2$ are both powers of 2? No, not at all.) If the hidden number were between 1 and 31, we would guess the middle number once again. The first guess would be 16, shrinking the number of integers in the interval from 31 to 15. The second guess would shrink the interval to seven integers; the third guess, to three integers; the fourth guess, to one integer; and the fifth guess would find the answer.

What if the initial interval had been [1, 63], 63 being one smaller than $2^6 = 64$? The numbers of integers in the intervals would have been 63, 31, 15, 7, 3, 1; six guesses would have sufficed.

In general, let's say the hidden number is known to be between, say, 1 and n (where n is 15, 31, and 63 above). The calculations are easier if n is one less than a power of two, so let's cheat –and it *is* cheating– to assume this. Please trust me that a messier version of the same argument goes through even when n is not one less than a power of two.

Suppose $n = 2^h - 1$ where h is a positive integer. We will first prove that h guesses suffice in this case. First I'll give an informal argument, which is not a proof.

Here it is. Initially the hidden number is between 1 and n. The first guess, the middle number, is $(n + 1)/2$, which is $2^h/2 = 2^{h-1}$. Suppose that that guess were wrong (for otherwise, one guess suffices, and $1 \leq h$). Then both remaining intervals have $(n - 1)/2 = 2^{h-1} - 1$ integers. The second guess decreases the number of integers in the interval to $((2^{h-1} - 1) - 1)/2 = 2^{h-2} - 1$. The third decreases the number of integers in the interval to $((2^{h-2} - 1) - 1)/2 = 2^{h-3} - 1$. You see the pattern. After $h - 1$ steps, the interval will contain $2^{h-(h-1)} - 1 = 2^1 - 1 = 2 - 1 = 1$ integer, and we will be done with one more query.

This is a fine argument, but it is not a proof, because of the need for "You see the pattern." Here's a formal proof by induction. (See Chap. 1 for an introduction to induction.)

Theorem 7.1 *Let h be any positive integer. Then h queries suffice to guess the missing number if it is known to be in any interval containing exactly $2^h - 1$ integers.*

Proof We will do induction on h. First we have to prove that the statement with $h = 1$ is true. But $2^1 - 1 = 1$, so the statement is that one query suffices if the number is known to lie in an interval containing only one integer, which is clearly true.

Now let h be any positive integer. Assume that for $l = 1, 2, 3, \ldots, h$, if the hidden number is known to lie in an interval containing exactly $2^l - 1$ integers, then l queries suffice. In particular, taking $l = h$, if the hidden number lies in an interval containing exactly $2^h - 1$ integers, then h queries suffice. We must show that if the hidden number lies in an interval containing exactly $2^{h+1} - 1$ integers, then $h + 1$ queries suffice.

Suppose that the hidden number lies in an interval containing exactly $2^{h+1} - 1$ integers. Ask if the hidden number is the middle integer.

If the answer is "Got it!," then one query sufficed; since $1 \leq h + 1$, we are done.

If the answer is "Lower," then the new interval contains exactly $[(2^{h+1} - 1) - 1]/2 = 2^h - 1$ integers. By the inductive assertion, h more queries suffice. Together with the one query already made, we will have made $h + 1$ queries, so $h + 1$ will have sufficed.

If the answer is "Higher," then, similarly, the new interval contains exactly $2^h - 1$ integers. As in the case of "Lower," h more queries suffice, for a total of $h + 1$. ∎

This means we're half done. I claimed this algorithm was optimal. *Optimal* means no other algorithm is better. It really *is* optimal, for every value of n. But to make the proof easier and less messy, I will prove that it's optimal only for values of n of the form $2^h - 1$ for a positive integer h. Since h queries are made by the binary-search algorithm, we must prove that no algorithm can always make at most $h - 1$ queries.

There is a subtle point here. I'm not claiming that there is no algorithm that *sometimes* makes fewer than h queries. For example, if the hidden number happens to be the first midpoint itself (which is $(n + 1)/2$), then the one query will suffice. I am claiming that there's no algorithm that *always* makes fewer than h queries.

Theorem 7.2 *There is no algorithm that always makes h − 1 or fewer queries.*

Proof This is the fun part. So far, we have focused on how to choose which numbers to query. Let's call this algorithm the *query-chooser* or sometimes just the *algorithm*. Now, we're going to switch sides, and play the role of the *query-answerer*, the one that says either "Got it!", "Higher," or "Lower." This person will attempt to make the algorithm ask as many queries as possible. Since fundamentally this "person" is trying to frustrate the query-chooser, we'll call this person the *adversary*, which is just a fancy word for "enemy."

This adversary keeps track of the (smallest) interval, as I described above, which necessarily contains the hidden number. Initially, of course, this interval is $[1, n]$. As time progresses, the interval shrinks in size.

Here's how the adversary answers the algorithm's queries. First of all, if the queried number is to the left of the interval, the adversary answers "Higher." If the queried number is to the right of the interval, the adversary answers "Lower." However, these would both be stupid (and wasteful) questions for an algorithm to ask. You see, the algorithm can *also* maintain the interval that the adversary is maintaining, and if it knows that the hidden number is in the interval, it automatically knows that the hidden number is higher than any guess which is below the interval, and lower than any guess which is higher than the interval, and why would it waste a guess on a query to which it already knows the answer? Since no self-respecting algorithm would ever query a number outside the current interval, let's assume that no such queries ever take place.

So let's tell the adversary how to answer a query to a number which is *inside* the interval. First of all, if the interval has only one integer, then the adversary replies, "Got it!", and the game is over. Otherwise, the adversary answers according to which side of the query is larger: if there are more integers to the left of the query than to the right, the adversary answers "Lower," and if there are more integers to the right of the query than to the left, then the adversary answers "Higher," and if there's a tie, the adversary can give either answer.

There's an important point here, which is that the adversary is supposed to have a hidden number in mind, and to answer the questions in a way consistent with the hidden number. What happened to the hidden number? This is a point that bedeviled me when I first saw it. What is going on? The algorithm's trying to find a hidden number, known by the adversary, and yet the adversary is answering questions "Got it!", "Higher," or "Lower," without having a hidden number? Isn't this wrong?

It turns out that it's not wrong. The adversary can claim, if questioned, that he or she had *any* one of the numbers in the interval in mind all along! For example, suppose $n = 15$ so that the initial interval was $[1, 15]$. Suppose that the algorithm queried 8, the middle number, first, and that the answer was "Higher." This shrinks the interval from $[1, 15]$ to $[9, 15]$. Then suppose the algorithm queried 12, the middle number, and that the algorithm replied "Lower." Now the interval has shrunken from $[9, 15]$ to $[9, 11]$. Both the algorithm and the adversary know that the hidden number is 9, 10, or 11. The key point is that the adversary can *pretend* that the hidden number is any of 9, 10, 11. The answers given by the adversary so far are consistent with *all* of these possibilities. No one can call the adversary a liar if he or she claims the hidden number was 9, or was 10, or was 11, for all of his or her answers to date are consistent with all three.

Personally, I find this a very intriguing point. The game is about finding a missing number, yet the adversary doesn't even need to have a hidden number in mind. How strange!

Here's another important point. As we know, at any time, there is an interval which is known to contain the "hidden number." *The algorithm cannot stop and declare it knows the hidden number until the interval contains only one number*, at which point one more query must be made (to which the answer will be "Got it!"). If the interval contains two or more numbers, there is no chance the adversary will respond "Got it!".

Let's think about how the interval changes size over time. Initially, the interval is $[1, n]$, which contains n integers. After a query, the adversary always says "Lower" or "Higher" according to which side of the query is larger. Suppose that the current interval contains s points. One of them is the query point, leaving $s - 1$ others. Since the two "sides" of the query together contain $s - 1$ points, one of the two sides contains at least $(s - 1)/2$ points. For example, if $n = 15$, at the beginning, $s = 15$. Excluding the query point, one of the two sides contains at least seven integers. If the original interval is $[1, n]$ where $n = 2^h - 1$, then the next interval contains at least $(n - 1)/2 = (2^h - 2)/2 = 2^{h-1} - 1$ integers. The one after that contains at least $[(2^{h-1} - 1) - 1]/2 = 2^{h-2} - 1$ integers. The one after *that* contains at least $[(2^{h-2} - 1) - 1]/2 = 2^{h-3} - 1$ points. After some number, say, t, queries, the interval contains at least $2^{h-t} - 1$ points. (This is easy to see but really requires an inductive proof which I omit.) Remember, the game continues until the interval has exactly one point, at which time one more query is necessary. Since just before the end the interval contains 1 point, we must have $2^{h-t} - 1 \le 1$, or $2^{h-t} \le 2$, or $h - t \le 1$, or $t \ge h - 1$. So at least $h - 1$ queries are necessary to get the interval down to one point, at which time one more query is necessary, or a total of $(h - 1) + 1 = h$ queries.

We have shown that h queries are necessary when $n = 2^h - 1$. ∎

7.2   Computers, Calculators, and Finite Precision

In Chap. 4 we discussed the square root of a number and showed, interestingly, that the square root of 2 is irrational. This is a fascinating fact. Notwithstanding this fact, from a computational standpoint, sometimes you need a square root. Now computers only represent decimal numbers of fixed precision, so surely cannot represent exactly $\sqrt{2}$. In fact, you don't have to go to square roots to find numbers that computers cannot represent exactly. Computers cannot even represent $1/3$

exactly, since its decimal form has an infinite string of digits.[1] If you want to have fun,[2] find a cheap calculator and compute $1/3$ and then multiply the result by 3. We all know that $(1/3) \cdot 3 = 1$. There is a good chance that the cheap calculator will display something like 0.999999999. The reason is that $1/3$ is *approximated* by 0.333333333; multiplying that by 3 gives 0.999999999. A good calculator, on the other hand, will maintain an *internal precision* which exceeds its *display precision*. What do I mean exactly? My old Texas Instruments SR-50A calculates all results accurately to 13 digits but rounds the results to 10 digits before displaying them. For example, when you ask for $1/3$, the calculator calculates $1/3$ to 13 digits: .3333333333333 but only displays the first 10: .3333333333. Then when you multiply by 3, it multiplies the *more accurate* version, .3333333333333 by 3, getting .9999999999999 and rounds it to 10 digits to display the result: hence it displays 1. Internally, the number is not 1, but what you see is $(1/3) \cdot 3 = 1$.

(As an unrelated aside, take your cheap dollar-store calculator and ask it to do $2 \cdot 3 + 4 \cdot 5$. The correct answer, of course, is $(2 \cdot 3) + (4 \cdot 5) = 6 + 20 = 26$. Your cheap calculator is likely to do $2 \cdot 3 = 6, 6 + 4 = 10, 10 \cdot 5 = 50$ and mistakenly give you 50, showing that it doesn't know about precedence of operations.)

Getting back to the discussion of precision, any calculator or computer can only calculate a decimal approximation of $\sqrt{2}$ (or of $1/3$) which is accurate to a certain number of decimal digits, perhaps 16, of accuracy. We cannot, of course, hope to calculate all the infinitely many digits of $\sqrt{2}$. But this leads to a very interesting question: how does one approximate square roots on a computer? How would you use addition, subtraction, multiplication, and division to approximate square root? Use binary search!

7.3   Square Roots via Binary Search

Let's suppose we want $\sqrt{c}$, where $c > 1$. (To get $\sqrt{c}$ where $c < 1$, see Exercise 7.1.)

The algorithm we'll use is based on, but not identical to, the binary search you've seen already. We will maintain an interval $I = [low, high]$ such that $\sqrt{c}$ is certain to be in I. It will always be true that $low \leq \sqrt{c} \leq high$. Initially, since $c > 1$, we can use $I = [1, c]$; certainly $1 < \sqrt{c} < c$ (because $1 < c$). If you know any smaller interval containing $\sqrt{c}$, you can use that instead.

We will seek an approximate square root which is within ϵ of the true square root, where ϵ is a user-defined tolerance, like $1e - 10$ (this is computer-speak for 10^{-10}). That strange-looking letter ϵ I just used is a Greek letter called "epsilon" and pronounced like "epsilon" and often used in mathematics to represent a small positive number.

---

[1] I will pretend that computers internally use decimal representations, since decimal representations are more familiar, rather than the binary representations they actually use.

[2] This is my idea of fun.

Following the idea of binary search, here's the algorithm. We will again use the absolute value of x, written $|x|$, which is x or $-x$, whichever is nonnegative. Absolute values are often used in measuring errors. We will use it to measure how far a number is from where it's supposed to be.

Here is the algorithm:

1. Let $\epsilon = 1e - 10$.
2. Let $low, high$ be any numbers such that $low \leq \sqrt{c} \leq high$. Since $c > 1$, one can take $low = 1$ and $high = c$, though a smaller interval would work better, especially when c is huge.
3. Repeat forever:
   (a) Let $midpoint = (low + high)/2$.
   (b) Let $errorbound = high - low$.
   (c) If $errorbound \leq \epsilon$, print the fact that the approximate square root is $midpoint$ and stop.
   (d) Otherwise:
      (i) Let $square = midpoint \cdot midpoint$.
      (ii) If $square = c$, print the fact that the square root is $midpoint$ and halt.
      (iii) If $square > c$, then $midpoint$ is too large to be the square root. Replace the pair $(low, high)$ by the pair $(low, midpoint)$. That is, set $high = midpoint$.
      (iv) Otherwise, $midpoint$ is too small to be the square root. Replace the pair $(low, high)$ by the pair $(midpoint, high)$. That is, set $low = midpoint$.

I've coded up this algorithm in Python; you can find the actual code in `binarysearchforsquareroot.py` on the website. Try running it for different values of c, some, like 9, with integer square roots, and some, like 2, without integer square roots. Each pass through the while loop is called an *iteration*. To *iterate* means to *repeat*. The code needs 51 iterations to find $\sqrt{2}$ to within a tolerance of 1e-15.

7.3.1 Time Analysis

Here's an analysis of the running time. Δ is a capital Greek letter *Delta* pronounced as it is written in English.

For any real x, we write $\lceil x \rceil$, read "the ceiling of x," to mean the smallest integer which is at least x. Hence $\lceil 5.01 \rceil = 6$ and $\lceil -7.99 \rceil = -7$.

Theorem 7.3 *Let Δ denote the starting value of $high - low$. Then the number of iterations needed to get an estimate of $\sqrt{c}$ which is within ϵ of the true square root is at most $\lceil \log_2(\Delta/\epsilon) \rceil$.*

Proof Each iteration through the loop cuts $high - low$ in half. At any time, the absolute value of the difference between the midpoint of the interval and the true square root is at most the length of the current interval. (Actually, it's at most half the length of the interval.) After i iterations, the length of the interval is exactly $\Delta/2^i$. Hence we are guaranteed to have the midpoint within distance ϵ of the true square root when $\Delta/2^i \le \epsilon$, i.e., $2^i \ge \Delta/\epsilon$, that is, $i \ge \log_2(\Delta/\epsilon)$, so i is the least integer greater than or equal to $\log_2(\Delta/\epsilon)$, that is, $i = \lceil \log_2(\Delta/\epsilon) \rceil$. If we use $\Delta = c - 1$, then the number of iterations is bounded by $i = \lceil \log_2((c-1)/\epsilon) \rceil$. $\blacksquare$

In the next chapter, we will see a different algorithm, not based on binary search, for finding square roots.

7.3.2 But Floating-Point Calculations Are Inexact!

I have been assuming that the calculations done by the computer are all exact, which we know is not true. Every computer stores its floating-point numbers (what mathematicians call "reals") with only finite precision. We saw this above when multiplying one-third by three. A consequence of this fact is that floating-point arithmetic is inexact. Things don't always turn out on a real computer as an analysis which assumes exact arithmetic would predict.

For example, if one tries to compute the square root of 2 with $\epsilon = 10^{-16}$, instead of 10^{-15} for which it runs beautifully, the binary-search code will run forever and never find a square root. However, there are alternatives. One can, instead, stop iterating when the relative error between *consecutive* midpoints is small enough and output the most recent midpoint.

7.4   Puzzle

Assume for this problem that the chance that a baby is a boy is exactly $1/2$.

At one point the government of China, to limit the growth of the population, instituted a "one family, one child" policy. It was illegal for families to have more than one child.

Think about what would have happened if the policy had been instead, "one family, one girl." That is, families could have an unlimited number of boys, but as soon as they had a girl, they would have had to stop having children. Under such a policy, a family could have zero boys, or one boy, or two boys, or three boys, or …20 boys (in addition to possibly having a girl). No family, however, could ever have two girls.[3]

Under such a policy, in the long run, what would be the fraction of boys and girls in China?

---

[3] I'd be out of luck, since I have three girls.

7.5   Exercises

Exercise 7.1 How can one compute the square root of an arbitrary positive real, given an algorithm for computing square roots only of numbers exceeding 1?

Exercise 7.2 Show how to use binary search to find a root of $f(x)$ where $f(x)$ is increasing.

Exercise 7.3 You are given an array a of length n of integers whose values increase for a while, reach a peak in a_h, and then decrease for the remainder of the array. Precisely, $a_0 < a_1 < a_2 < \cdots < a_h > a_{h+1} > a_{h+2} > \cdots > a_{n-1}$. The value of h is unknown. How can you find h in approximately $c \log_2(n)$ queries, for some small constant c?

Exercise 7.4 * Prove that $n - 1$ comparisons are necessary to find the largest number in a list a of n distinct numbers, if all an algorithm can do is compare elements in the input list.

Exercise 7.5 *

(a) Recall that $n!$, read "n factorial," equals $1 \cdot 2 \cdot 3 \cdots n$.
   Prove that at least $\log_2(n!)$ comparisons are necessary to sort a list a of n distinct numbers, if all an algorithm can do is compare two elements of the list.
(b) Show that $\log_2(n!) \geq (1/4)n \log_2 n$ for all $n \geq 4$.

Exercise 7.6 Binary search shrinks the size of the problem by at least half in each iteration. In this problem, we discuss a different algorithm, *QuickSelect*, which shrinks its problem size by roughly half in each iteration.
   QuickSelect solves the following problem. You are given an unsorted array a of n distinct reals and a positive integer $k \in \{1, 2, 3, \ldots, n\}$. Your goal is to find the kth largest number in a ($k = 1$ means the largest, $k = n$ means the smallest). QuickSelect is *randomized* in the sense that it makes random choices. Here's how *QuickSelect*(a, k) works; some details are omitted for you to fill in:

1. Pick a random element z in a. This is done by picking a random $i \in \{0, 1, 2, 3, \ldots, n - 1\}$ and setting z to be the ith element of a.
2. Compare every element in a to z to find the rank l of z in a. (The *rank* of z is l if there are exactly $l - 1$ elements of a bigger than z and exactly $n - l$ elements smaller.) Put the $l - 1$ elements of a which are larger than z into an unsorted array L and put the $n - l$ elements of a which are smaller than z into an unsorted array S. Now there are three cases:

   - If $l = k$, then we are done; output z and halt.

- If $l > k$, then the number we are looking for is the kth largest element of L. Recursively calculate $x = QuickSelect(L, k)$ and output x.
- Otherwise, $l < k$. The number we are looking for is the rth largest element of S (where you have to figure out r). Recursively calculate $x = QuickSelect(S, r)$ and output x.

(a) Find r in terms of k and the size of L so that the rth largest element of S is the kth largest element in a.

(b) Because z is chosen uniformly at random from a, very roughly, L and S will have average length approximately $n/2$. Assuming incorrectly that L and S always have size exactly half the size of a (until the algorithm terminates), and that n is a power of two, prove the following facts:

- The number of iterations is at most $1 + \log_2 n$.
- The sum of the lengths of all the arrays encountered is exactly $2n - 1$.

If instead the analysis is done carefully and correctly, one sees that finding the kth largest using $QuickSelect$ takes expected time bounded by Cn (for some constant C) and is hence much faster than finding the kth largest by sorting, since sorting by comparisons takes time at least $(1/4)n \log_2 n$ according to Exercise 7.5(b). For large enough n, $(1/4)n \log_2 n$ is much greater than Cn (regardless of C).

7.5.1 Hints

7.4 Say an element a_i of the list is a *loser* if the algorithm compared a_i to some other element a_j and discovered that $a_j > a_i$. Each comparison can only create one loser. Argue that if there are $n - 2$ or fewer losers at the end, then the algorithm can't know with certainty which element is the largest.

7.5

(a) There are $n!$ possible permutations of n numbers. Initially, all $n!$ permutations are possible. After i comparisons have been made, suppose c_i permutations of the numbers are consistent with the results of the i comparisons. The algorithm cannot know the sorted order until $c_i = 1$. Furthermore, when a_r is compared with a_s, one of the two outcomes ($a_r < a_s$ or $a_r > a_s$) causes c_i to drop by at most a factor of two. An adversary can choose to answer each query so that c_i drops the least.

(b) Use $n! \geq (n/2)^{n/2}$ for $n \geq 1$ and $\log_2 (n/2) \geq (1/2) \log_2 n$ if $n \geq 4$.

Newton's Method

8

8.1   Introduction

No doubt you've heard of Isaac Newton, the "discoverer" of gravity. Prior to Newton's "discovery" of gravity, everyone and everything just floated around. It's good that Newton floated along to discover gravity, since without him, everything on Earth which is not attached to the Earth, including our atmosphere, would have floated off into outer space. This means we can thank Newton for our lives.

More seriously, Newton didn't discover gravity—he quantified it: he wrote down the equations that govern gravity. You may have heard the apocryphal story about his being hit on the head by an apple falling from an apple tree. Maybe this happened, maybe it did not, but Newton's key observation was that the same force that caused the apple to fall caused the moon to remain in orbit around the earth. That was an epiphany for the ages.

Newton did not just quantify gravity. He introduced the basic laws of mechanics, the branch of physics dealing with motion. He co-invented calculus. He contributed to optics and invented the reflecting telescope. He was certainly one of the greatest scientists of all time.

In this chapter we will discuss a technique attributed to Newton to find zeros of functions. A *zero* or *root* of a function is pretty much what you'd expect it to be: it's a value of x which makes the function equal to 0. For example, if $f(x) = x^2 - 4$, then both 2 and -2 are zeros of the function, since $2^2 - 4 = 0$ and $(-2)^2 - 4 = 0$. Root-finding is a crucial operation. If you want to calculate the length of a footpath running along the diagonal of a rectangular field, you need to use the Pythagorean Theorem and find a square root; that's root-finding. Root-finding is pervasive in engineering and physics. Root-finding is needed in optimization, the mathematical field of study that studies how to do things optimally, e.g., most quickly or most cheaply.

© The Author(s), under exclusive license to Springer Nature Switzerland AG 2023
H. Karloff, *Mathematical Thinking*, Compact Textbooks in Mathematics,
https://doi.org/10.1007/978-3-031-33203-6_8

What do I mean by optimization? I'll give you a personal anecdote. About 10 years ago, I worked at Yahoo Labs, before Yahoo was bought by Verizon, and later by Apollo Global Management. An Internet company, Yahoo made much of its revenue by placing ads on websites. There were many different Yahoo sites, among them Yahoo Finance, Yahoo Sports, Yahoo Mail, and Yahoo Screen. Obviously Yahoo wanted to keep its advertisers happy... so they would keep advertising.

Advertisers would come to Yahoo and say something like this: "We spent one million dollars on Yahoo ads over the last quarter. We liked the results enough to want to spend two million dollars on Yahoo ads this quarter. But how should we allocate the $2M in advertising among the various Yahoo sites to get the most conversions?" (Purchases are known as "conversions" in the online world.)

Yahoo Labs had designed a model to predict, for a given large advertiser, given a number of dollars spent (the *spend*) on a given Yahoo site, how many conversions would result. This function was nonlinear, since advertising twice as much on a site doesn't win you twice as many conversions. The model was based on machine learning (in case you know what that is).

The question was, how should the advertiser split its $2M across the many Yahoo sites to get the maximum number of conversions in total? How much on Yahoo Finance? How much on Yahoo Sports? How much on Yahoo Mail? How much on Yahoo Screen? This was the optimization problem I worked on.

Unfortunately, one problem we faced in practice was that advertisers often didn't trust Yahoo. We would run our optimizer to find the best allocation of advertising spend *for the advertiser* and often the advertiser would ask, "How do I know you're not finding the allocation of advertising spend that's best *for Yahoo*?" Their concern was that we were pushing them to advertise on unpopular Yahoo sites, the ones other advertisers didn't want to advertise on. The truth is that we never did that. We always optimized for the advertiser, but often they didn't trust us.

Now you know what optimization is.

I will describe Newton's method as a method for computing square roots.

You may have been tormented by the question as to how the ancient Babylonians computed square roots. I'm glad you asked. There is evidence that the ancient Babylonians, living in southern Mesopotamia (which is modern-day Iraq) in the first third of the second millennium BCE, more than 3700 years ago, maybe even more than 4000 years ago, used the following simple method to compute the square root of a positive number c:

Start with any positive approximation a to $\sqrt{c}$, for example, $a = c$. Then compute the average of a and c/a and use that as the next approximation to $\sqrt{c}$. Repeat.

For example, if $c = 2$ and a initially is 2, then the next a will be the average of 2 and $2/2$, or $(2 + 2/2)/2 = 1.5$. The next a will be the average of 1.5 and $2/1.5$, or $(1.5 + 2/1.5)/2 = 3/4 + 2/3 = 17/12 = 1.41666666....$ The third a would be the average of $17/12$ and $2/(17/12) = 24/17$, or $17/24 + 12/17 = 577/408 = 1.414215686....$ The fourth would be $(577/408 + 2/(577/408))/2 = 1.414213562375....$ Given that $\sqrt{2} = 1.414213562373...,$ this is an amazingly good approximation. The first discrepancy is in the 12th digit after the decimal point!

A variant of this method is used by today's calculators and computers to calculate square roots.

It appears that the Babylonians only did the first iteration of this process. People say that the Babylonians wrote in cuneiform on clay tablets (though I imagine that possibly they did most of their work on lined $8 - 1/2 \times 11$ paper but only the clay tablets survived till today). They worked in base 60, meaning that they had 60 digits in their *sexagesimal* number system, as compared with the 10 digits in our *decimal* number system. Lest you malign their use of base 60, remember that we, in a sense, use base 60 for minutes and seconds. After all, the modern time 00:10:23 (shortly after midnight) is naturally interpreted as a base-60 number, as there are 60 minutes in every hour and 60 seconds in every minute.

Since 60 is such a round number, one can express half an hour (30 minutes), a third of an hour (20 minutes), a quarter of an hour (15 minutes), a fifth of an hour (12 minutes), a sixth (10 minutes), a tenth (6 minutes), even a twelfth of an hour (5 minutes), and so on, all without needing fractions. Try doing that with a decimal system of time in which an hour has 10 "minutes."

By the way, the Babylonians knew the Pythagorean Theorem (see Theorem 4.8); maybe the need to compute lengths of hypotenuses was one reason they needed to calculate square roots. Interestingly, to multiply two positive integers a and b, instead of using a multiplication table (which would have had to go up to 59×59 and have 3481 entries), they used a table of perfect squares up to $59^2 = 3481$ and a formula similar to $ab = [(a + b)^2 - (a - b)^2]/4$. For example, $27 \cdot 17 = (44^2 - 10^2)/4 = (1936 - 100)/4 = 1836/4 = 459$. What an interesting way to multiply by hand!

I have no idea if Newton was aware of the Babylonian method for computing square roots. Newton came up with a method for solving equations which includes the Babylonian method but which is much more general. I will show how Newton's method leads to the Babylonian method. And we will analyze the quality of the estimates that Newton's method gives.

Like the binary search algorithm for square roots, Newton's method is iterative, but Newton's method converges much faster than the binary search algorithm to the square root.

Suppose we want to compute $\sqrt{c}$ where $c > 0$. Consider the function $y = x^2 - c$. Figure 8.1 is a plot of $y = x^2 - c$ when $c = 2$. The important point here is that the zeros of $y = x^2 - c$ are those x's with $x^2 - c = 0$, that is, x's whose square is c.

8.2   How to Improve a Solution

When you're working with iterative algorithms, the key question is, "Given a solution, how can I improve it?" If I start with one estimate a for the square root of c, how can I generate a better estimate? And then a better one after that?

Here is Newton's idea. Take the curve $y = x^2 - c$, the zeros of which are the two square roots of c. Consider that point $P = (a, a^2 - c)$ which the curve passes through. Let's *approximate* the curve at P by a straight line. Which line should we

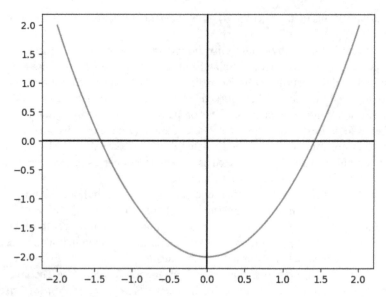

Fig. 8.1 A plot of $y = x^2 - 2$ vs x

pick? It seems natural to pick the line which passes through P and which has the same slope at P. The line passing through P with the same slope as the curve is called the *tangent line at P*.

Have you ever seen slope defined for a curve? If not, no problem. The slope of the curve $y = x^2 - c$ at the point $P = (a, a^2 - c)$ with x-coordinate a is exactly $2a$. I'll explain where this comes from in Sect. 8.4, but for now, please just believe me.

According to Newton's idea, we should approximate the curve at point P by the line L passing through $P = (a, a^2 - c)$ having slope $2a$.

For example, see Fig. 8.2 to see the tangent line to $y = x^2 - 2$ (so $c = 2$) at $a = 2$.

At this point probably Newton said to himself, "If we're approximating the curve by the line L, then a good guess for the zero of the curve is the zero of the line." Where does the line cross the x-axis? Let's call that point $(a', 0)$. The slope of the line L is $2a$ and also, by definition of slope, the slope of the line segment connecting P and the point $(a', 0)$. Recall that the slope of a line passing through two points is *(change in y)/(change in x)*. In this case, the first point is $(a', 0)$ and the second point is $(a, a^2 - c)$. The change in y is $(a^2 - c) - 0$ and the change in x is $a - a'$. This makes the slope $((a^2 - c) - 0)/(a - a')$. Since we know the slope is $2a$, we have $(a^2 - c)/(a - a') = 2a$ and we then have $(a^2 - c)/(2a) = a - a'$. Hence a' satisfies

$$(2a)a' - (a^2 + c) = 0.$$

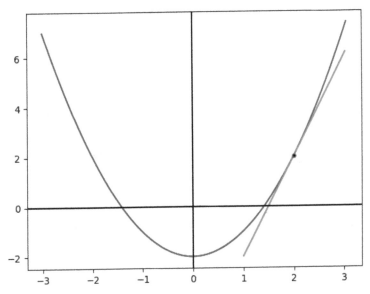

Fig. 8.2 A plot of $y = x^2 - 2$ vs x with the tangent line at $x = 2$ drawn in red. Notice that the intersection of the tangent line with the x-axis closely approximates the positive root of $x^2 - 2$, that is, $\sqrt{2}$

Solving this equation for a' gives us

$$a' = \frac{a^2 + c}{2a} = \frac{a + c/a}{2}.$$

Given an estimate a, the next estimate should be $a' = (1/2)(a + c/a)$, the average of a and c/a. This is precisely the Babylonian method!

If a is smaller than $\sqrt{c}$, then c/a is larger, and if a is larger than $\sqrt{c}$, then c/a is smaller. In both cases the next estimate a' of $\sqrt{c}$ is the average of two estimates of $\sqrt{c}$, one too small and one too large.

I've coded up the algorithm in Python—see `newtonsmethodforsquare root.py`—and you can see that's it's as simple as the binary search algorithm. However, now instead of taking 51 iterations to find $\sqrt{2}$, it only needs five. We'll see why later, but trust me, Newton's method is an amazing way to find square roots.

8.3   Extension to Other Functions

And it's not great only for square roots. You can try the method to find zeros of other functions, provided that you know how to find the tangent line at P. However, in general there's no guarantee that it will work, since there's no guarantee that the sequence of estimates will converge to a zero of the function.

If you think about it, there really isn't anything very special about the function $y = x^2 - c$ to which we're applying Newton's method. So let's try the same idea for another function we'll just call $f(x)$. What did we do so far? Recall that the value of a function at point a is written $f(a)$; above, $f(a)$ was $a^2 - c$. Let's write the slope of a function f at a point a as $s(a)$; above, $s(a)$ was $2a$. If a' will be our improved estimate, P is the point $(a, f(a))$, and L is the line of slope $s(a)$ through P, then the slope of L is $(f(a) - 0)/(a - a')$, which must equal $s(a)$: $f(a)/(a - a') = s(a)$, so

$$a' = a - f(a)/s(a).$$

(Above, $a' = a - (a^2 - c)/(2a)$.) This is how Newton's method applies to more general functions.

It is interesting to note that while Newton applied his method to polynomials, even the great Newton himself missed the generalization given here. Only in 1740, 55 years after Newton's method was originally published, was the generalization given here published (by Thomas Simpson).

8.4   Tangent Lines

This section is a bit complicated, and you can skip it if you want to. But read on if you want to see why I said the slope of $y = x^2 - c$ at $x = a$ is $2a$.

If the equation of a line is $y = mx + b$, then the slope is m. For example, the slope of $y = 4x - 17$ is 4. The slope is nothing more than the change in y divided by the change in x.

Let's study the example $y = x^2$. For lines, we could measure *change in y* and *change in x* between any two points. For curves like $y = x^2$, we have to specify between which two points we're calculating *change in y* and *change in x*. To measure the slope "at the point $x = a$," let's take a positive b and look at two points, one at $x = a - b$ (to the left of a), where $y = (a - b)^2$, and one at $x = a + b$ (to the right of a), where $y = (a + b)^2$. The change in y is

$$(a + b)^2 - (a - b)^2 = (a^2 + 2ab + b^2) - (a^2 - 2ab + b^2) = 4ab.$$

(Coincidentally, this is the formula the Babylonians used for multiplication.) The change in x is $(a + b) - (a - b) = 2b$. Hence the slope is $4ab/(2b) = 2a$.[1] For this reason, we define the *slope* of $y = x^2$ at the point $x = a$ to be $2a$. Nothing changes, by the way, if, instead of computing the slope of $y = x^2$, we compute the slope of

---

[1] It is a somewhat magical fact that this ratio doesn't depend on b in this case. It might have, in which we'd have had to see what happens to the ratio for supersmall values of b, such as 0 itself.

$y = x^2 - 1$, or $y = x^2 - 2$, or $y = x^2 - 4$, etc., so the slope of each of these at the point $x = a$ is also $2a$.

Think of the tangent line through a point as the line which best approximates the function at that point.

8.5   How Good Is Newton's Method?

Is Newton's method any good? In fact, it's astoundingly good... but what does this mean? The big issue is whether the next estimate is much better than the current one.

Let's assume that the number c whose square root we want is at least 1. Let's also assume that our estimate a for the square root of c is an *overestimate*, that is, it's either equal to $\sqrt{c}$ or larger. What can we say about the next estimate—might it be an underestimate? It turns out that it can't be and that useful fact is easy to prove.

Lemma 8.1 *If c is at least 1 and a is at least $\sqrt{c}$, and $a' = (a + c/a)/2$, then a' is at least $\sqrt{c}$.*

Proof We clearly have $(a - c/a)^2 \geq 0$, since any number squared is nonnegative. Hence

$$a^2 - 2 \cdot a \cdot c/a + c^2/a^2 = a^2 - 2c + c^2/a^2 \geq 0.$$

So, adding $4c$ to both sides, we have

$$a^2 + 2c + \frac{c^2}{a^2} \geq 4c.$$

The left-hand side is $(a + c/a)^2$. Hence

$$\left(a + \frac{c}{a}\right)^2 \geq 4c,$$

and hence, taking the square root of both sides and then dividing by 2,

$$\frac{a + c/a}{2} \geq \sqrt{c}.$$

The left-hand side is a'.   ■

We have shown that regardless of the value of the estimate a, provided it's at least $\sqrt{c}$, the next estimate will also be at least $\sqrt{c}$. And then the one after that will be at least $\sqrt{c}$. And so will the next one. And so on. Hence all the estimates will be at least $\sqrt{c}$.

Let us define the *error e* of an estimate a which is at least $\sqrt{c}$ to be $e = a - \sqrt{c}$ (which cannot be negative).

Let's start by proving a relationship between the error of one estimate and the error of the next one.

Theorem 8.1 *Suppose $a \geq \sqrt{c} \geq 1$ and let its error e be $a - \sqrt{c}$. Let $a' = (a + c/a)/2$ and let its error e' be $a' - \sqrt{c}$. Then*

$$e' = \frac{e^2}{2a}.$$

The proof is a few lines of algebra.

Proof Now $e^2 = (a - \sqrt{c})(a - \sqrt{c}) = a^2 - 2a\sqrt{c} + c$. Recall that $a' = (1/2)(a + c/a)$. Hence

$$\frac{e^2}{2a} = \frac{c - 2a\sqrt{c} + a^2}{2a}$$

$$= -\sqrt{c} + \frac{1}{2}\left(\frac{c}{a} + a\right)$$

$$= -\sqrt{c} + a'$$

$$= a' - \sqrt{c}$$

$$= e'.$$

∎

Is this good or bad? Let's start instead by showing something else, that $e' \leq (1/2)e$ always.

Corollary 8.1 $e' \leq e/2$ *for all i.*

Proof We have $e = a - \sqrt{c} \leq a$ (because $\sqrt{c} \geq 0$). So $e \leq a$. Multiply both sides by $e/(2a)$, which is positive: $e^2/(2a) \leq a \cdot (e/(2a)) = e/2$. But the left-hand side, by Theorem 8.1, equals e'. ∎

We have proven that the error shrinks by at least a factor of 2 in each iteration. This is akin to the binary-search algorithm for square root. (In that case, there is no guarantee that each error is half of the previous one, though.)

This is a nice result already. However, Newton's algorithm doesn't just decrease the error by a factor of two. Once the error gets down to about $1/2$, it starts shrinking superrapidly. Since Theorem 8.1 shows that $e' \leq e^2$ if the denominator $2a$ is at least 1, which it always is if $c \geq 1$, we have:

if the error now is at most $\frac{1}{2}$, then the following error will be at most $(\frac{1}{2})^2 = \frac{1}{4}$; which implies that the following error will be at most $(\frac{1}{4})^2 = \frac{1}{16}$; which implies that the next error will be at most $(\frac{1}{16})^2 = \frac{1}{256}$; and which implies that the next error will be at most $(\frac{1}{256})^2 = \frac{1}{65536}$; and this implies that the following error will be at most $(\frac{1}{65536})^2 = \frac{1}{4294967296}$.

For example, here are the estimates to $\sqrt{4}$ (which you know to be 2) produced by Newton's method: $4.0, 2.5, 2.05, 2.000609756097561, 2.0000000929222947, 2.000000000000002$. This is called *quadratic convergence*. So Newton's algorithm combines dividing-by-2 convergence (known as *linear convergence*) with quadratic convergence, which really shines once the error drops below $1/2$.

If the number c whose square root you want is small, say, between 1 and 4, the quadratic convergence will kick in quickly, but if it's very large or very small, you might end up waiting a long time. What should one do, patiently wait as the error drops repeatedly only by factors of two, till quadratic convergence kicks in? No. A better idea is to divide c by some very large or very small power 4^k of four (k being positive or negative) to adjust the number to a number $c' = c/4^k$ between 1 and 4, find an approximation x' to $\sqrt{c'}$, and then multiply x' by 2^k to get an estimate x of $\sqrt{c}$. Because computers work in base 2 (binary), multiplying and dividing by powers of 4 or 2 is superfast.

Here's a warning, however. Floating-point arithmetic is inexact, and the analysis given above assumes the arithmetic is exact. If you ask for too small an error relative to the number c whose square root you want, say, you try to estimate $\sqrt{10^{100}}$ (10^{100} having been called a *googol* long before Google existed) with error at most 1.0, the arithmetic will be too imprecise, and the errors will not decrease as the theory predicts they should. In fact, the code may never terminate.

8.6   A Personal Note

Years ago I interviewed for a job with Facebook in Menlo Park, California. The first question I was asked was, "How can you compute square roots on a computer without a square root operation?"

I thought I was fully prepared for this question, since a friend of mine had interviewed with Facebook about 2 years earlier and had been asked exactly this question (and he had told me so). Not only that, I had taken numerical analysis as an undergraduate where I had learned Newton's method.

In order to be honest, I immediately told the interviewer that a friend of mine had told me that he'd been asked this question in his Facebook interview and that since I had taken numerical analysis, I knew Newton's method. I tried to be polite.

I didn't get the job, and heard later that they thought I was arrogant. I guess I wasn't polite enough.

8.7   Puzzle

Assume the earth is a perfect sphere.

A hunter left his home one day, walked precisely one kilometer south and then one kilometer east, and shot a bear. He dragged the bear one kilometer north to his home. What color was the bear? More interestingly, find the set S of all of points on the surface of the earth where the hunter's home could have been.

This is one of my favorite puzzles. Do not answer it quickly. If you think you have the answer, prove that (1) every point in S satisfies the description of the home, and (2) no point outside of S satisfies the description.

8.8   Exercises

Exercise 8.1 Prove that the slope of $f(x) = x^3$ at $x = a$ is $3a^2$. Do this by computing

$$\frac{(a+b)^3 - (a-b)^3}{2b}$$

for supersmall values of b (like 0).

Exercise 8.2 Using the result of Exercise 8.1, give the formula for Newton's method applied to $f(x) = x^3 - c$ for c a real.

Exercise 8.3 Suppose $f(x)$ is a function.[2]

(a) Show informally that if the slope of $f(x)$ at $x = a$ is positive, then f is increasing at $x = a$ (meaning that the values of f at points slightly larger than a are larger than $f(a)$ and that the values of f at points slightly smaller than a are smaller than $f(a)$).
(b) Similarly, show informally that if the slope of $f(x)$ at $x = a$ is negative, then f is decreasing at $x = a$ (meaning that the values of f at points slightly larger than a are smaller than $f(a)$ and that the values of f at points slightly smaller than a are larger than $f(a)$).
(c) Show that if $x = a$ is a point that has the largest value $f(x)$ over all x's, that is, $f(a) = \max_x f(x)$, then the slope of f at $x = a$ is 0.

Exercise 8.4 * You own a widget-producing factory. Your revenue $r(x)$ in dollars if you produce x widgets is $r(x) = 1000x - x^2$, for $0 \leq x \leq 1000$. Your cost $c(x)$ in dollars if you produce x widgets is $100x$. How many widgets should you manufacture to maximize your profit, which is $r(x) - c(x)$?

---

[2] for which a "slope" is defined

Exercise 8.5

(a) Write a Python program which finds a root of a function $f(x)$ using Newton's method. Your program should call a method f which returns the value of $f(x)$ at $x = a$ and a method $fprime$ which returns the slope of $f(x)$ at $x = a$. Your program will also take a starting value for the iteration.

(b) Use your program to find the roots of $f(x) = x^2 - 4x + 3$, which are 1 and 3. Run your program starting at 3.1 and 1.1.

Your program should converge to 3 when started at 3.1 and to 1 when started at 1.1. You now know that when there are multiple roots, which one your program converges to depends on the point at which it started.

Exercise 8.6 Use your program from Exercise 8.5 to find a root of $f(x) = x \cdot \ln x - 3$. Here $\ln x = \log_e(x)$ and is written math.log(x) in Python. (I don't know how to find a root of this equation without a computer.) The slope of $f(x)$ at the point $x = a$ is $1 + \ln(a)$. Give your answer rounded to ten digits after the decimal point.

8.8.1 Hint

8.4 Use Exercise 8.3.

Graph Theory

9

9.1 Introduction

What's a "graph?" No doubt you've seen graphs of y vs x, as in a plot of $y = 2x + 3$ or $y = 3x^2 + 5$. This kind of graph is exactly what I'm *not* talking about. You can forget about $y = \sin(2x)$, $z = 3x + 2y - 4$, and even $y = 2\tan(3x^2)$.

We're going to talk in this chapter about a different kind of "graph," the kind with "vertices" (often called "nodes") and "edges." Specifically, a *graph* is a finite non-empty set V, the *vertex set*, of *vertices*, and a set E, the *edge set*, of unordered pairs, called *edges*, of distinct nodes. Such a graph is usually written as (V, E). Edges are often written by writing the two vertices in the edge next to each other with a hyphen between them, remembering that reversing the order would leave the edge completely unchanged.

I remember when I saw this, vertices were so abstract I wondered what vertices really were. The answer is, any objects around.

This is so general a definition it seems almost meaningless. For example, here's a graph: $V = \{New\ York, Los\ Angeles, Chicago, baseball, grapefruit\}$, with the edge set $E = \{New\ York - baseball, grapefruit - Chicago\}$. This (silly) graph, which we'll call S for "silly," has five vertices and two edges. You may have seen sets before. The defining properties of a set, as you probably know, are that a set's elements are all different and that the order in which the elements appear is irrelevant. This means that $\{New\ York, New\ York, Chicago\}$ is exactly the same set as $\{New\ York, Chicago\}$ and the same set as $\{Chicago, New\ York\}$. Because the edge set E consists of *unordered* edges, each edge is the same whether it's written forward or backward. This means that $New\ York - baseball$ is *exactly* the same edge as $baseball - New\ York$.

© The Author(s), under exclusive license to Springer Nature Switzerland AG 2023
H. Karloff, *Mathematical Thinking*, Compact Textbooks in Mathematics,
https://doi.org/10.1007/978-3-031-33203-6_9

Fig. 9.1 A drawing of K_4
with a crossing

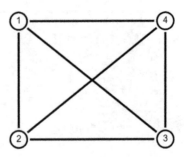

Here is a whole "family" of graphs. Let n be a positive integer, let $V =$ $\{1, 2, 3, ..., n\}$ and let E be the set of *all* possible edges. This graph is called K_n, the n-vertex *complete graph*, sometimes called a *clique*. For example, K_4 has $V = \{1, 2, 3, 4\}$ and $E = \{1 - 2, 1 - 3, 1 - 4, 2 - 3, 2 - 4, 3 - 4\}$. No graph with n vertices has more edges than the complete graph.

Sometimes people allow there to be more than one edge between a single pair of vertices. For example, we could take K_4 but have *two* copies of the edge between vertices 4 and 2. This cannot happen in a graph, so we'll call such objects *multigraphs* instead of "graphs."

The *degree* of a vertex is the number of edges touching it. Degrees will be very important in the Bridges of Königsberg puzzle. For example, in K_n, the complete graph on n vertices, every vertex has degree $n - 1$.

Often it's useful to attach numbers to the edges of a graph. When each edge comes with a number, the graph is called a *weighted* graph. Often the number on an edge is called the *length*, *weight*, or *cost* of the edge.

One nice thing about working with graphs is that they're very concrete, unlike the situation in some other branches of mathematics. You don't have to worry about infinities. You can draw your graphs on a sheet of paper—this makes visualizing graphs a piece of cake. You draw each vertex as a dot and draw an edge between vertices x and y as a curve between the dot for x and the dot for y. For example, in Fig. 9.1 is a picture of K_4.

What are graphs used for? Why do people bother with them? It turns out that graphs are really useful. Edges represent relationships between vertices. There are many interesting questions one can ask about graphs. Here is one example.

9.2   Planarity

As soon as you start drawing graphs on paper, you start wondering if those annoying edge crossings have to exist. It would be aesthetically pleasing if the curves representing two different edges didn't touch at all, except possibly at their endpoints. (The *endpoints* of edge $u - v$ are the vertices u and v.)

Given a graph, can it be drawn on a sheet of paper with the edges not touching, except possibly at the endpoints? Since the sheet of paper represents a Euclidean plane, such graphs are called *planar*. Now when I first saw this definition, I was

Fig. 9.2 A drawing of K_4 with no crossings. While this graph is drawn with straight line segments for edges, curves are also allowed

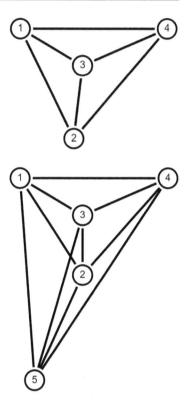

Fig. 9.3 A drawing of K_5 with one crossing. The edges $1-2$ and $3-5$ cross, but no other pair crosses. It is impossible to draw K_5 on paper, even with curved edges, with no crossings

confused. Some graphs seem to be both planar and nonplanar. Consider the complete graph K_4 I mentioned above. It can be drawn on the plane in many different ways. There is one drawing above, in which some two edges do cross. However, it can also be drawn as in Fig. 9.2. It's important to understand that the *graphs*, which consist of a vertex set and an edge set, are either planar or nonplanar, not *drawings* of graphs. K_4 can be drawn in many different ways. There happens to be a drawing (Fig. 9.2) in which no edges cross, so K_4 is planar.

Try K_5, the complete graph on five vertices. In Fig. 9.3 you'll see a drawing of K_5 with one pair of edges crossing. It turns out that this is the minimum number of crossings. There is simply no way to draw K_5 on a plane with no crossing edges; in other words, K_5 is nonplanar. This is a deep statement!

Which other graphs are nonplanar? Clearly K_6, K_7, K_8, and so on are nonplanar, because all of them contain a K_5, and if you could draw them without crossing edges, then you could also draw K_5 without crossing edges, and you can't. In fact, we are using the fact here that if you can find a "copy" of a nonplanar graph H "inside" a bigger graph G, then G cannot be planar itself. (I am leaving "copy" and "inside" undefined, but I hope you know what I mean.)

But here's another nonplanar graph: $K_{3,3}$. $K_{3,3}$ is a graph with six vertices, three named *boys*, b_1, b_2, b_3, and three named *girls*, g_1, g_2, g_3, and all nine possible edges

Fig. 9.4 A drawing of $K_{3,3}$
with many crossings

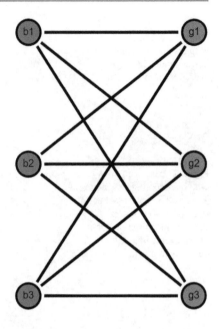

between a girl and a boy, and no other edges. In other words, the vertex set V is
$\{b_1, b_2, b_3, g_1, g_2, g_3\}$, and the edge set E is $\{b_1 - g_1, b_1 - g_2, b_1 - g_3, b_2 - g_1, b_2 -
g_2, b_2 - g_3, b_3 - g_1, b_3 - g_2, b_3 - g_3\}$. Figure 9.4 is a drawing of $K_{3,3}$ with many
crossing edges. In Fig. 9.5 you'll see a drawing of $K_{3,3}$ with only one crossing.

Is $K_{3,3}$ planar or not? It is a very nontrivial fact that $K_{3,3}$ is nonplanar: no matter
how hard you try, you'll never manage to draw $K_{3,3}$ on a sheet of paper, even with
curved edges, without crossing edges. Go ahead. Try.

We've seen above that if a graph H is nonplanar, then adding additional vertices
or edges will leave it nonplanar. It turns out that there's another way to generate
a new nonplanar graph from a nonplanar graph H: by "subdividing" edges. To
subdivide an edge $u - v$ is to replace that one edge by a path of vertices. I haven't
told you what a "path" is, but here's an example. In K_5, which has vertex set
$V = \{1, 2, 3, 4, 5\}$, with all ten possible edges, replace the edge $1 - 3$ by the path
$1 - x, x - y, y - z, z - 3$, where x, y, z are three new vertices. This new graph,
with $5 + 3 = 8$ vertices, does *not* contain K_5, in fact. By *contains* K_5 I mean that
you can remove zero or more nodes and zero or more edges and be left with a copy
of K_5. Make sure you understand this point. It almost does, as instead of the edge
$1 - 3$ it contains a path, but it does not contain K_5. However, this new graph is
also nonplanar. Why? Because if we could draw the new graph in the plane without
crossing edges, then we could draw K_5 without crossing edges, simply by erasing
the "dots" representing the new vertices x, y, z, and we know that we cannot draw
K_5 in the plane without crossing edges.

If you start with a nonplanar graph H and subdivide one edge, you get a
nonplanar graph. By the same logic, if you start with a nonplanar graph H, and

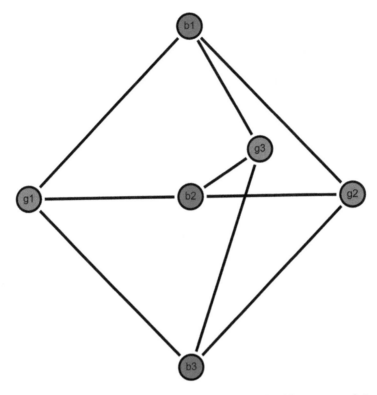

Fig. 9.5 A drawing of $K_{3,3}$ with one crossing. The edges $b_2 - g_2$ and $b_3 - g_3$ cross, but no other pair cross. It is impossible to draw $K_{3,3}$ on paper, even with curved edges, with no crossings

subdivide one *or more* edges, the final graph will not be planar, either. The final graph, called a *subdivision* of H, will be nonplanar.

We have learned already that:

1. Any subdivision of K_5 is nonplanar.
2. Any subdivision of $K_{3,3}$ is nonplanar.

Furthermore, any graph that contains a nonplanar graph is itself nonplanar. It follows that any graph that contains a subdivision of K_5 or a subdivision of $K_{3,3}$ is nonplanar.

Now we reach the remarkable fact: these are the *only* nonplanar graphs! Here is the theorem, which is not easy to prove (and we will not do so).

Theorem 9.1 *Kuratowski's Theorem. A graph G is planar if, and only if, it does not contain a subdivision of K_5 or $K_{3,3}$. In other words, if a graph contains a subdivision of K_5 or $K_{3,3}$, then it is nonplanar. Otherwise, it is planar.*

Just to be clear, don't miss the words "contain" and "contains." Graphs which are not themselves subdivisions of K_5 or $K_{3,3}$ but which *contain* subdivisions of K_5 or $K_{3,3}$ are also nonplanar.

In my opinion, this is a remarkable fact. Once you prove that K_5 and $K_{3,3}$ are nonplanar, it is immediately obvious that any graph containing a subdivision of K_5 or $K_{3,3}$ must be nonplanar. The truly amazing fact is that these are the *only* nonplanar graphs!

It turns out that there are fast computer algorithms that will take a graph G and tell you whether or not G is planar. Interestingly, the algorithms do not directly work by looking for subdivisions of K_5 or $K_{3,3}$.

9.3   Graph Coloring

9.3.1   Vertex Coloring

Two vertices u, v in a graph are *adjacent* if there is an edge between u and v. Remember that $u - v$ and $v - u$ refer to the same edge, so if u, v are adjacent, then v, u are adjacent. If u and v are adjacent, then v is a *neighbor* of u and u is a *neighbor* of v.

One of the fun things you can do with graphs is to color their vertices such that adjacent vertices get different colors. By "color" I could mean the colors you are used to, like red, blue, and green, but since it's convenient to have an unbounded palette of colors, we will use positive integers as "colors." The important question when coloring vertices is always, given a graph G, what is the minimum number of colors needed to color its vertices? The minimum number of colors needed to color a graph G is called its *chromatic number* and denoted $\chi(G)$, where that strange symbol is a Greek letter "chi" pronounced "kigh" (and rhyming with "high").

The hardest n-vertex graphs to color are the complete graphs K_n containing all possible edges.

Theorem 9.2 $\chi(K_n) = n$.

Proof Clearly one can safely give all n vertices n different colors, so $\chi(G) \leq n$. But in fact, one *must* give all n vertices n different colors, since every pair of vertices is adjacent, so $\chi(G) \geq n$. Hence $\chi(G) = n$. ∎

Let $\Delta(G)$ denote the maximum degree in graph G.

Theorem 9.3 $\chi(G) \leq \Delta(G) + 1$.

Proof Here is a very simple algorithm for coloring a graph with at most $\Delta(G) + 1$ colors. Label the, say, n, vertices in graph G $v_1, v_2, v_3, ..., v_n$ in any order. Use a palette $\{1, 2, 3, ..., \Delta(G) + 1\}$, start from vertex v_1, and proceed in order. When you reach a vertex v_i, some subset of v_i's neighbors have already been colored.

However, v_i has at most $\Delta(G)$ neighbors, so even if they were all colored with different colors, they could use only $\Delta(G)$ of the $\Delta(G) + 1$ colors in the palette. Use one of the unused colors in the palette to color v_i. ∎

How about coloring planar graphs? Why would anyone want to color planar graphs? Well, cartographers (who make maps) want adjacent countries of maps to be of different colors. After all, how would you know where the USA ends and Mexico begins if they're both the same color? If you build a graph with one vertex per country, and an edge between two countries that share a border, like the USA and Mexico, you get a planar graph.[1] Now you see why people were interested in coloring planar graphs.

Some graphs require a huge number of colors. How about planar graphs? Does Theorem 9.3 give a good bound on the number of colors needed for planar graphs? Unfortunately, it does not, since some planar graphs have extremely high maximum degree. Consider the n-vertex *star* graph S_n consisting of one vertex adjacent to $n - 1$ neighbors, with no other edges; S_n is planar and $\Delta(S_n) = n - 1$. Theorem 9.3 says that $\chi(S_n) \leq n$, which is true but uninteresting; in fact $\chi(S_n) = 2$.

Is there a single number C such that all planar graphs, regardless of how many vertices they have, can be colored with at most C colors? Somewhat remarkably, there is! One of the most famous theorems from mathematics, the *Four-Color Theorem* states that every planar graph can be colored with at most four colors.

Theorem 9.4 *The Four-Color Theorem. Every planar graph can be colored with at most four colors.*

This remarkable theorem has its own remarkable history. It was first conjectured in 1852 and then "proven" in 1879 by Kempe. The "proof" was believed until 1890 when Heawood pointed out a hole in the "proof." Then, for approximately 86 years, it was known as the *Four-Color Conjecture* until, in 1977, finally a correct proof was published by Appel and Haken. This remarkable proof was computer-assisted and, indeed, was the first time a famous theorem was proven with extensive assistance— in this case, 1200 hours of computer time—by a computer. Since there are infinitely many planar graphs, the computer was not naively used to color planar graphs, because it would never have finished. Instead, roughly speaking, the computer was used to find an *unavoidable* set of 1936 "configurations" and to show that each configuration in the unavoidable set was "reducible." Here, again speaking roughly, "unavoidable" means that every planar graph must have one of the configurations and "reducible" means, using induction, that a graph containing a configuration could be reduced to a smaller graph.

---

[1] This is true if each country is a contiguous region. If a country contains regions like Alaska that are not contiguous with the rest of the country, and all the regions of a country must be the same color, then the graph might not be planar.

Even the Appel-Haken proof contained some mistakes, but those mistakes have all been fixed. Truly, the history of coloring planar graphs is itself colorful.

Since the proof of the Four-Color Theorem is so complicated, I will prove instead (while relying on one unproven fact) that every planar graph can be colored with at most six colors. There is an important fact about planar graphs that simplifies the coloring of planar graphs. It is not that hard to prove, but I will state it without proof.

Theorem 9.5 *A planar graph on n vertices has at most 3n − 3 edges.*

By contrast, a complete graph has $n(n-1)/2$ edges—this is an exercise—which is far more. Using the theorem, it is an easy exercise to prove this theorem.

Theorem 9.6 *Every planar graph has a vertex of degree at most five.*

Here is our theorem.

Theorem 9.7 *Any planar graph G can be colored with at most six colors.*

Proof Suppose G has n vertices. Rename the vertices $v_1, v_2, v_2, ..., v_n$ in the following way. By Theorem 9.6, G has a vertex of degree at most five. Call that vertex v_n.

Now remove v_n from G and all edges touching it. The remaining graph is also planar (do you see why?), so has a vertex of degree at most five. Let v_{n-1} be that vertex.

Now remove v_{n-1} from that graph and all edges touching it. The remaining graph is also planar, so has a vertex of degree at most five. Let v_{n-2} be that vertex.

Continue in this way, always choosing a vertex of degree at most five in the graph that remains. In this way, generate a sequence $< v_1, v_2, v_3, ..., v_{n-2}, v_{n-1}, v_n >$ of vertices. Each vertex in this list has at most five edges to preceding vertices.

Now color the graph using palette $\{1, 2, 3, 4, 5, 6\}$ in the forward order $< v_1, v_2, v_3, ..., v_n >$. When it comes time to color vertex v_i, at most five of its neighbors are already colored, so a palette of size six suffices. ∎

It is not hard to improve this theorem to get down to five colors.

You might be asking yourself, "Every planar graph can be 4-colored, but can one find the coloring quickly on a computer?" The answer is yes, though this is not obvious.

You might also ask yourself, "Can every planar graph be 3-colored?" Of course not: K_4, which is planar, cannot be three-colored.

So now you ask yourself, "Can one determine quickly if a given planar graph can be 3-colored?" The surprising answer is that this problem is known to be *NP-Complete*. This is a fancy term that means, in a formal sense, that the 3-COLORING OF PLANAR GRAPHS problem is the most difficult problem in a huge class of problems, and so that people believe, but cannot prove, that there is no fast algorithm for the problem.

By contrast, there is an easy algorithm to determine if a graph, planar or not, can be two-colored. This is an exercise.

9.3.2 Edge Coloring

Graphs have zero or one edge between a given pair of vertices. Recall that in a *multigraph* G, there can be multiple edges between a single pair of vertices. The maximum number of edges between any single pair of vertices is the *multiplicity* $\mu(G)$ (μ being a Greek letter pronounced "myoo") of the multigraph. The multiplicity of a graph is one, if there are any edges at all.

The *degree* of a vertex in a multigraph is the number of edges touching the vertex, as in the case of graphs, with $\Delta(G)$ denoting the maximum degree, as in graphs.

We can color the *edges* of a multigraph, in such a way that edges that share an endpoint must get different colors. The minimum number of colors needed to edge color a multigraph (or graph) G is called its *chromatic index* and denoted $\chi'(G)$ (which is read "chi prime of G").

You might be asking, since we're edge coloring multigraphs, why we never bothered to vertex color multigraphs. The reason is that once there is an edge between vertices u and v, u and v must get different colors. Additional edges between u and v would change nothing.

Vizing proved the following theorem.

Theorem 9.8 *Vizing's Theorem. For any multigraph G of maximum degree $\Delta(G)$ and multiplicity $\mu(G)$,*

$$\Delta(G) \leq \chi'(G) \leq \Delta(G) + \mu(G).$$

Unfortunately Vizing's Theorem is not so easy to prove so I won't give a proof.

Vizing's Theorem is remarkable because it tells you, for a graph, almost exactly what $\chi'(G)$ is.

Corollary 9.1 *For a graph G, $\Delta(G) \leq \chi'(G) \leq \Delta(G) + 1$.*

Notice the contrast with Theorem 9.3, for vertex coloring. There, all we could say was that $\chi(G)$ was between 2 and $\Delta(G) + 1$. By contrast, here we see that $\chi'(G)$ for a graph G must be either $\Delta(G)$ or $\Delta(G) + 1$. This is amazing.

Proof $\Delta(G) \leq \chi'(G)$ is obvious, because all the edges touching the vertex of degree $\Delta(G)$ must get different colors. The other inequality follows from Vizing's Theorem and the fact that $\mu(G) \leq 1$ in a graph. ∎

9.4   Eulerian Multigraphs

Now let's take a walk.

A *walk* in a graph is a finite sequence of vertices, each one adjacent to the next one. For example, if $a - b$ and $b - c$ are edges, then $< a, b, c, b, c, b, a >$ is a walk; as you can see, repeated vertices are allowed. A walk in which no *vertex* appears twice is a *path*. To *traverse* an edge is to walk across it. A walk in which no *edge* is traversed twice (in either direction) is a *trail*. If the trail's first and last vertices are the same, that is, if it's *closed*, then it's called a *tour*.

In deference to mathematician Leonhard Euler (pronounced, believe it or not, as "oiler"), the founder of graph theory, an *Euler trail* of a graph is a trail which passes through each edge (in some direction) exactly once. An Euler trail may pass through a vertex zero times (if no edges touch that vertex), one time, or multiple times. For example, in the complete graph K_5 on $V = \{1, 2, 3, 4, 5\}$, here is an Euler trail: $< 1, 2, 3, 4, 5, 2, 4, 1, 5, 3, 1 >$. An *Euler tour* is an Euler trail which is a tour (i.e., which starts and ends at the same vertex). That last example is also an Euler tour.

Here is an important puzzle. Consider the multigraph K shown in Fig. 9.6. There are four vertices 1, 2, 3, 4, and edges 4–1, 4–2, 4–3, 1–3, 1–3, 2–3, 2–3. (It's a multigraph because it has repeated edges, such as 1–3.) Technically we should change the definitions above, which were intended for graphs, which don't have repeated edges, so that they make sense for multigraphs, but since I think you understand, I won't bother. Is there an Euler trail in this multigraph? Is there an Euler tour? (Since every tour is a trail, if there is an Euler tour, then there is an Euler trail.)

This puzzle is important because it was the origin of graph theory. There was a city known as Königsberg in a region called Prussia (the precursor to Germany), which no longer exists. The same city today is known as Kaliningrad and is in Russia. There were seven bridges in the city, represented by edges in the multigraph

Fig. 9.6 The Königsberg
bridges multigraph

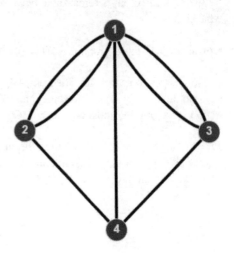

K ("K" for Königsberg), over a river, which connected four regions of the city, represented by vertices in the multigraph. The residents of the city gave themselves this challenge. Starting from any one of the regions, was it possible to walk across each bridge exactly once? (The challenge did not require them to end in the same region from which they started.) Was it possible? In today's language, is there an Euler trail in Fig. 9.6? Notice that since they did not hope to end at the same vertex from which they started, they were seeking an Euler trail, not an Euler tour.

In 1736, Leonhard Euler showed that the challenge was impossible. His elegant solution is viewed as the first result of graph theory. By the way, in reality, the real challenge to the residents of Königsberg involved walking on the roads and bridges of the city. To abstract away the details of the city and replace the challenge by one involving walking on a multigraph, instead of on roads and bridges, was part of Euler's genius.

Theorem 9.9   *There is no Euler trail in K.*

Can you prove that it is impossible to find an Euler trail in K? Try yourself before reading the proof.

Proof Remember that the degree of a vertex is the number of edges touching it. In multigraph K, vertices 2, 3, and 4 have degree 3 and vertex 1 has degree 5. Hence there are a total of four vertices having odd degree.

Now suppose, for a contradiction, that there is a walk W in K traversing each edge exactly once, that is, an Euler trail. That walk W must start at some vertex, u, and end at some vertex, v. Possibly u and v are the same vertex, possibly not.

Here's the key observation: every time the walk enters and exits a vertex z, it "uses up" two of the edges touching z.

Hence every vertex *other than u and v* must have even degree (because W traverses each edge exactly once). The only vertices of odd degree (if there are any) can be u and v. If $u \neq v$, there can be at most two vertices of odd degree (namely, u and v), whereas if $u = v$, then u (and v) must have even degree (because the path leaves u at the beginning and enters it at the end), so there must be 0 vertices of odd degree.

However, K has four vertices of odd degree. It follows that there is no Euler trail in K.   ■

A multigraph having an Euler tour is known as *Eulerian*. By analogy with the theorem above, we have this theorem.

Theorem 9.10   *If multigraph G has an Euler trail, then the number of odd-degree vertices is 0 or 2. If multigraph G has an Euler tour, then the number of odd-degree vertices is 0, that is, every vertex has even degree.*

Now if you've been doing math long enough, as I have, when you see a theorem like this one, you ask about its converse. This is the statement you get by swapping the two sides of the "If-then." In this case, the converse of the second statement is,

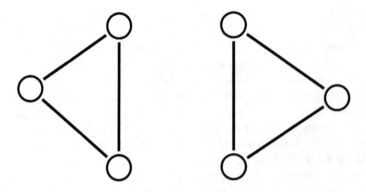

Fig. 9.7 Two triangles

Conjecture 9.1 If every vertex of multigraph G has even degree, then G has an Euler tour.

We call something which might be true but about which we're not sure a *conjecture*.

Now there is no reason why the conjecture should be true. A statement and its converse are two different statements. Quite possibly one is true and not the other. In this case, in fact, the conjecture is false. Consider the (multi)graph in Fig. 9.7 consisting of two disjoint triangles with no edges between them. Every vertex has even degree, yet clearly there is no Euler tour. So the converse is false, but it's false for some trivial reason: the (multi)graph isn't even "connected," in other words, there's no way to get from one triangle to the other, so how could there possibly be an Euler tour?

Let's talk about connectedness for a moment. A pair u, v of vertices in a multigraph is *connected* if there is a walk between them. A multigraph is itself *connected* if every pair of vertices in the multigraph is connected. For example, K_4 and K_5 are connected, but the graph in Fig. 9.7 consisting of two disjoint triangles is not.

An *isolated vertex* is a vertex of degree 0.

We have this simple theorem.

Theorem 9.11 *If G is an Eulerian multigraph having no isolated vertices, then G is connected.*

Proof Because there are no isolated vertices, every vertex touches at least one edge. The Euler tour traverses every edge and hence visits every vertex. Hence the Euler tour contains a $u - v$ walk for any pair u, v of vertices.   ∎

Since isolated vertices aren't very interesting, and multigraphs without isolated vertices have to be connected to be Eulerian, from now on we will only look for Euler tours in connected multigraphs.

Now it seems natural to make this revised conjecture.

Conjecture 9.2 If G is a connected multigraph in which every vertex has even degree, then G is Eulerian.

It is not obvious whether the conjecture is true or not. We know that every connected Eulerian multigraph has only vertices of even degree, but we have no idea whether every connected multigraph which has only vertices of even degree is Eulerian. *These are two entirely different statements.*

In fact, this latest conjecture *is* true, but it is not obvious. We need to give an entirely new proof. Our proof will need a simple lemma. A *lemma* is a theorem used in the proof of another theorem. I remember when I first saw lemmas, I wondered, "What's the difference between a lemma and a theorem?" The answer is that formally there is no difference at all. The distinction is in how the result is used.

Let's upgrade the conjecture to a theorem.

Theorem 9.12 *If G is a connected multigraph in which every vertex has even degree, then G is Eulerian.*

The proof of the theorem will be *algorithmic*, which means that it gives an algorithm to construct the object (an Euler tour, in this case) that we're looking for. Sometimes in mathematics there are proofs that show that something (like an Euler tour) exists without giving an algorithm to construct it. These are often clever, indirect proofs. This is not one of those proofs. The present proof will show how to construct an Euler tour.

Before we get started, we need to prove a simple lemma.

Lemma 9.1 *Let H be a multigraph with at least one edge, in which every vertex has even degree. Let u be any node of H which is not isolated. Then one can find a trail in H which starts and ends at u and which has at least one edge.*

Proof Starting with u, take a walk through the multigraph as long as possible, as long as no edge ever gets traversed twice. Stop only when necessary. This walk is a trail with at least one edge because u is not isolated. It turns out that the trail, which started at u, must also end at u; this is not obvious. The reason is that each time the walk entered a vertex v other than u, it used up one edge; since the degree was even, there was at least one available edge from which to depart v. The process of visiting v used up two edges touching v, leaving the remaining degree even (but possibly zero). This means that the walk could not possibly have gotten "stuck" at any vertex other than u, so, since it cannot go on forever in a finite multigraph without repeating edges, it must have gotten "stuck" at u. A walk like this that starts and ends at the same node, and which never reuses an edge, is a closed trail.   ■

Notice that any closed trail uses up an even number of edges touching each node.

To prove Theorem 9.12, we are going to build a sequence $< W_1, W_2, W_3, ... >$ of longer and longer closed trails, the last one so long that it will contain every edge of G and hence be a closed Euler trail, that is, an Euler tour. This means that our goal is to take a closed trail W_i in G and, unless it already traverses all edges of G, to find a new closed trail W_{i+1} in G traversing more edges than W_i traversed.

Let's get started. Apply Lemma 9.1 to the connected multigraph G to construct a closed trail W_1 with at least one edge. W_1 might traverse all edges of G and it might not. If it does, then it's already an Euler tour of G so the proof is finished. Let's assume that it does not.

Unfortunately, we need another lemma.

Lemma 9.2 *Let W be a closed trail in a* connected *multigraph H. Suppose W does not traverse all edges of H. Then there is some edge $u - v$ not traversed by W such that u is visited by W.*

Notice that if H is the two-disjoint-triangles multigraph above and W, a closed trail on one of the triangles, then there is no edge $u - v$ not traversed by W with u visited by W, and the lemma doesn't hold in this case. The reason is that the two-disjoint-triangles multigraph is disconnected. This shows that the proof of Lemma 9.2 must use the fact that H is connected. *It is important when writing proofs to know which hypotheses must be used.*

Proof Let H be a connected multigraph with vertex set V and edge set E, and let W be a closed trail in H which does not traverse all edges of H. There are two cases: either W visits all vertices of H or it does not.

1. Suppose W visits all vertices of H.

   Since W doesn't traverse all edges of H, let $u - v$ be any edge it did not traverse. Since W visits all vertices of H, u is visited by W, and we are done: $u - v$ is the desired edge.

2. Suppose W does not visit all vertices of H. Let S be the set of vertices visited by W; S is a proper subset of the vertex set.

   Consider all edges of H, if any, with one endpoint in S and the other not in S. If there were no such edges, then there could be no walk in the multigraph between any vertex in S and any vertex not in S (and there is at least one vertex not in S). Since H is connected, there must be at least one edge $u - v$ with u in S and v not in S. This edge is the desired edge and we are done.   ■

Having done the groundwork, we are ready to show how to take one closed trail W_i and, from it, if it doesn't traverse all edges, to generate a longer closed trail W_{i+1}. That was our goal.

Given a connected multigraph G all of whose vertices have even degree, and a closed trail W_i in G which doesn't traverse all edges of G, let $u - v$ be an untraversed edge such that node u is visited by W_i. (Lemma 9.2 proves that such an edge exists.) Being a closed trail, W_i "uses up" an even number of edges at each vertex. Hence if we remove the traversed edges, every vertex in the remaining multigraph still will have even degree. Now, in the remaining set of edges, Lemma 9.1 implies that there is a trail Z, starting and ending at u, having at least one edge. We now have two closed trails: W_i, which passes through u, and Z, which starts and ends at u and uses none of the edges that W_i used. "Stitch together" the two trails to form one longer closed trail W_{i+1}, as follows. Take trail W_i. As soon as you get to vertex u the first time, detour through the trail Z (which starts and ends at u), following it around until it ends at u. Then finish the trail W_i. Call the new "combined" closed trail W_{i+1}. W_{i+1} is a closed trail which contains more edges than W_i did. This was our goal, since, if you remember, if we repeatedly generate longer and longer closed trails, the last one will traverse every edge of G and hence be an Euler tour. The proof is complete. ∎

9.5   Minimum Spanning Trees

9.5.1   Basic Facts

In this section we will discuss the problem of creating a cheapest connected graph given a bunch of vertices and weighted edges between them. For example, suppose you work for a telecommunications company, like ATT, for which I worked for many years. You might be interested in connecting a bunch of buildings so that they can communicate with each other. You'd like to connect them as cheaply as possible. You might in reality have a million buildings to connect, but to take a simple example, suppose you have six buildings, which we will call, with a total lack of imagination, h_1, h_2, h_3, h_4, h_5, and h_6. Suppose you are given the costs of building direct connections between certain pairs of buildings. You can see the costs of these edges in Fig. 9.8. These are two-way connections, so that if we build the edge $h_1 - h_2$, data will be able to flow from h_1 to h_2 and also from h_2 to h_1. Some edges, such as $h_1 - h_6$, are either impossible or too expensive to build. Let's assume that the cost of every edge is positive.

Your boss says to you, "Please tell me how to connect those six buildings in the cheapest possible way." You're lucky today, because here there are only six buildings to connect, but could you do it if there were one million buildings? You need a method that will work well, not only when there are six buildings, but when there are one million.

Let's think a little bit about what an optimal (i.e., best possible) solution would look like. When coming up with an algorithm to find an optimal solution, it often helps to think about what an optimal solution must look like. In this case, it helps to think about cycles.

Definition 9.1 A *cycle* is a walk on four or more vertices, all of which are distinct except that the first vertex and last vertex are the same. A cycle's *length* is the number of its edges.

For example, in Fig. 9.8, $< h_2, h_6, h_5, h_4, h_1, h_2 >$ is a cycle of length five, which we'll call C.

We need the concept of "subgraph."

Definition 9.2 A *subgraph* $H = (V', E')$ of a graph $G = (V, E)$ is a graph having V' a subset of V and E' a subset of E. That is, you get H from G by removing zero or more vertices (and all the edges touching those vertices) as well as zero or more of the edges that remain. The subgraph is called *spanning* if $V' = V$, that is, if no vertices were removed.

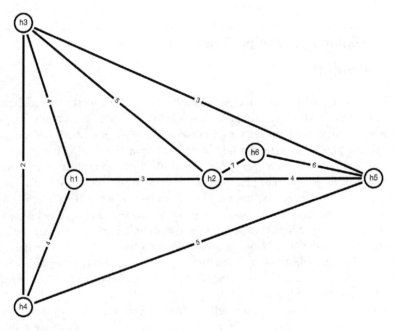

Fig. 9.8 Minimum spanning tree example

Lemma 9.3 *An optimal solution never contains a cycle. More precisely, suppose $G = (V, E)$ is a connected graph in which every edge has a positive cost. Let H be a spanning subgraph of G, which is connected, and which has minimum cost among all connected spanning subgraphs of G. Then H contains no cycle.*

For example, consider the graph drawn in Fig. 9.8. Suppose your friend says that the cheapest set of edges needed to connect the graph is to take cycle C plus edge $h_2 - h_3$, in other words, to take this set of edges: $\{h_2 - h_6, h_6 - h_5, h_5 - h_4, h_4 - h_1, h_1 - h_2, h_2 - h_3\}$. Could this set possibly be optimal?

Think for a moment and figure out why that set of edges couldn't possibly be optimal.

The key observation is that if you removed any one edge from the cycle C, leaving all other edges intact, the resulting graph would still be connected. The reason is that if a walk used the omitted edge, instead of the omitted edge one could use the rest of the cycle to go between the endpoints of the removed edge. For example, if a walk from some vertex u to some vertex v used edge $h_6 - h_5$, the part using that edge could be replaced by the walk $< h_6, h_2, h_1, h_4, h_5 >$, so there would still be a walk from u to v. Furthermore, the new graph would have one fewer edge, and the cost of that edge was positive. It follows that the new connected graph (the one without $h_6 - h_5$) would be cheaper than the old one, and hence the old one wasn't optimal.

This already contains the ideas of a proof, so I won't give a formal proof.

So we've learned something interesting—the cheapest connected spanning subgraph of a graph with positive-cost edges never has a cycle.

Definition 9.3 An acyclic graph (one without cycles) is called a *forest* and a connected acyclic graph is called a *tree*.

See Fig. 9.9 for an example of a forest.

Definition 9.4 A subgraph which is a tree and which contains all vertices of the parent graph is called a *spanning tree*. See Fig. 9.10 for an example of a graph G and a spanning tree of G.

It's easy to see that every connected graph has a spanning tree.

Theorem 9.13 *Every connected graph $G = (V, E)$ has a spanning tree.*

Proof Let $F = E$. Repeatedly find any cycle in the graph (V, F) and remove any one edge of the cycle from F; doing so leaves the graph (V, F) connected. Eventually, there are no cycles left, at which time the graph (V, F) is an acyclic, connected, and spanning subgraph of G, and hence a spanning tree of G. ∎

Here is an interesting fact about spanning trees of n-node graphs: they all have the same number of edges.

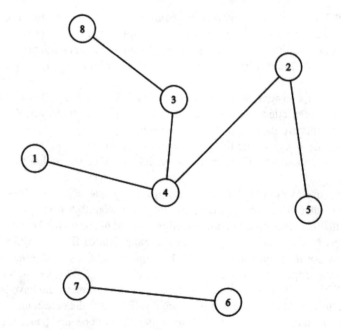

Fig. 9.9 A forest on eight nodes

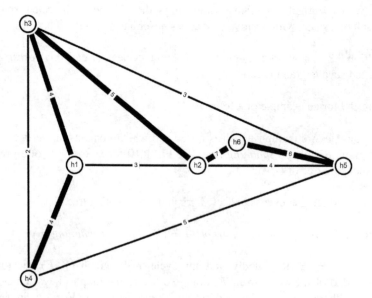

Fig. 9.10 A graph and a spanning tree, containing the thick edges, of the graph

Fig. 9.11 This graph has
three components, a triangle
on vertices 1, 3, and 7; a path
on vertices 5, 4, 2, and 8; and
an isolated vertex, vertex 6

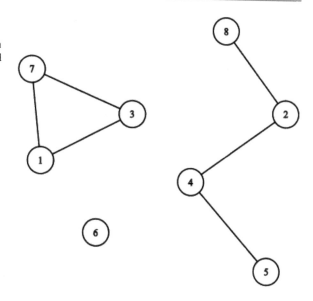

Theorem 9.14 *Any spanning tree in an n-node graph G has exactly n − 1 edges.*

For the proof, we need the notion of a "component" of a graph. Vertices u and
v in a graph $G = (V, E)$ are connected if there is a walk (or path) between them.
(There is a walk between two vertices if and only if there is a path between them,
so "connected" can be defined either way.) Now the "connected" relation is an
equivalence relation (see Exercise 9.2), so by Exercise 2.2, there is a partition of
the vertex set such that vertices u and v are connected if and only if they're in the
same part of the partition.

Definition 9.5 In a graph $G = (V, E)$, a *component* is the portion of the graph on
the vertices in one part of the partition.

See Fig. 9.11 for an example.
   Now we start the proof of Theorem 9.14.

Proof In G, suppose $T = (V, Z)$ is a spanning tree with $n = |V|$ vertices and
$m = |Z|$ edges. Our goal is to show that $m = n − 1$.
   Label the edges of Z arbitrarily as $e_1, e_2, ..., e_m$. Starting with a graph
$T_0 = (V, \varnothing)$ having no edges, add the edges one at a time in order. Let
$T_i = (V, \{e_1, e_2, ..., e_i\})$, $0 \leq i \leq m$, be the ith graph. Since $T_m = T = (V, Z)$ is
a spanning tree and hence acyclic, all of the T_i's, being subgraphs of T_m, are also
acyclic.
   Now think about how we got T_i from T_{i-1}. We did this by adding edge e_i to T_{i-1}.
Suppose e_i's endpoints are u and v. Now here's the main point of the proof: *in* T_{i-1},
u and v are in different components. For if u and v were in the same component,

there would be a walk (and hence also a path) from u to v in T_{i-1}. That path, together with e_i, would constitute a cycle, and hence T_i would have a cycle, but we saw above that all the T_i's were acyclic.

Now, we're almost done—we just have to count components. T_0, having n vertices and no edges, has n components. When we added edge e_i to T_{i-1}, its endpoints were in different components. This means that adding e_i merged the component containing u with the component containing v; therefore T_i has exactly one fewer component than T_{i-1} has.

Since T_0 has n components, T_1 has $n-1$; T_2 has $n-2$; T_3 has $n-3$; ...; and T_m has $n - m$. But T_m, being a spanning tree, is connected, and hence has one component. So $n - m = 1$ and $m = n - 1$. ∎

9.5.2   The Minimum Spanning Tree Algorithm

Assume every edge has a positive cost. Every connected graph G has some spanning tree, but how do you find a spanning tree of minimum cost? This spanning tree, known as a *minimum spanning tree*, is the one you would want if you really had to pay to construct edges. It's expensive to lay cable.

How would you find a minimum spanning tree in a weighted graph G? This is the main subject of this section. While the motivation for studying the problem is clear, another reason to study the problem exists: the algorithm is simple, natural, and beautiful. It is rare in the world of mathematics for the first algorithm you think of for a problem to actually be correct! In this case, the algorithm is *greedy*, a term loosely defined to mean that you repeatedly make decisions without looking ahead. I can't emphasize enough how unusual it is for a greedy algorithm to be optimal.

Let me give an example of a problem for which being greedy doesn't work, so you can see that greedy algorithms don't always work. Let's say you want to find the shortest walk from vertex s to vertex t. A shortest walk in a graph with positive edge weights will never use the same vertex twice, so a shortest walk is just a shortest path.

Suppose you have a map of four cities, not-so-creatively named 1, 2, 3, 4, and you want to find the shortest path from vertex 1 to vertex 4. There are four edges in the graph: 1–2, of cost 1; 1–3, of cost 2; 2–4, of cost 1,000,000; and 3–4, of cost 1. (See Fig. 9.12.) Now it's obvious that the shortest path from 1 to 4 goes through vertex 3 and has cost $2 + 1 = 3$. Keep in mind that finding shortest paths in 1000-node graphs isn't so easy, as just glancing at the graph won't work: you need an algorithm.

Now here's where being greedy is a very *bad* idea! Suppose you start at vertex 1. You decide to build one path iteratively by always appending to the end of the current path the cheapest edge available among those that go to a new vertex. Being greedy, you start by using the cheapest edge touching node 1, that is, the edge to node 2. Your next step would be to take the cheapest edge out of node 2, yet there

Fig. 9.12 A graph in which
we seek a shortest path from
vertex 1 to vertex 4

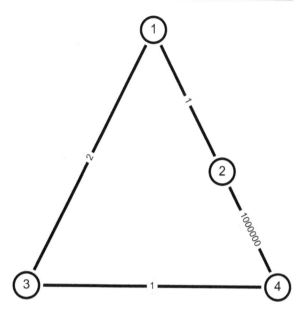

is only one other edge touching node 2: edge $2 - 4$, which has cost 1,000,000.
You've reached node 4 and can stop, but your total cost is 1,000,001; however,
the path through vertex 3 has length only 3. Behaving greedily has cost you a
lot. Finding shortest paths is often done with a non-greedy algorithm known as
Dijkstra's Algorithm, which you can read about on Wikipedia.

The remarkable thing about minimum spanning trees is that behaving greedily
actually *is* the best thing to do. We will have to prove this—it's far from obvious.

Enough said. What's the algorithm? Just how does one find a minimum spanning
tree?

Since we are going to be greedy, we start by sorting the edges into increasing
cost. (Technically, the correct term is "nondecreasing" since if two edges have the
same cost, the sorted list of costs won't be increasing.) Let's call the cheapest edge
e_1, the next cheapest e_2, the third cheapest e_3, etc. If two edges have the same cost—
that is, there's a tie—you can choose to place either one first. We're going to build
the minimum spanning tree by including some edges and excluding some others.
We will examine the edges in order, starting from the cheapest edge e_1. Remember
that minimum spanning trees never contain cycles. Now here's the key part: *add the
current edge if, and only if, it does not create a cycle with the edges chosen so far.*

That's the whole algorithm. Sort the edges into increasing order by cost. Scan
the edges in order, including an edge when you get to it if (and only if) adding it
would not form a cycle. At the end, you have a minimum spanning tree. What could
be simpler? Here is a formal description of the algorithm.

The Minimum Spanning Tree Algorithm

Let G be a connected graph with vertex set V and edge set E with, say, m edges.

Fig. 9.13 A weighted
version of K_4

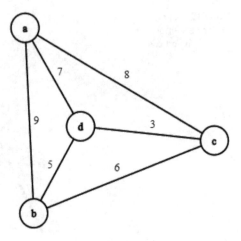

1. Sort the edges into increasing ordering by cost, so that $cost(e_1) \leq cost(e_2) \leq cost(e_3) \leq \cdots \leq cost(e_m)$.
2. Let F be an empty set.
3. For $i = 1, 2, 3, \ldots, m$, do:
   (a) Consider adding edge e_i to the graph (V, F). If adding e_i would *not* create a cycle, then add e_i to F. (And otherwise, don't add e_i to F.)

At the end, (V, F) will be a minimum spanning tree of G.

It's a remarkable fact that this algorithm works. I concede that it's a natural algorithm to try, but why should making such greedy, short-sighted decisions guarantee that the final tree has minimum cost among all spanning trees?

Consider the four-node complete graph K_4 with vertices a, b, c, d, instead of $1, 2, 3, 4$, and with the costs shown in Fig. 9.13. If you run the algorithm on this graph G, here's what happens. Initially F is empty. After sorting, we have edge e_1 is $c - d$; edge e_2 is $b - d$; edge e_3 is $b - c$; e_4 is $a - d$; e_5 is $a - c$; and e_6 is $a - b$.

Adding $e_1 = c - d$ to F, which is currently empty, would clearly create no cycle, so we choose $c - d$ and set $F = \{c - d\}$.

Adding $e_2 = b - d$ to F means that F would be $\{c - d, b - d\}$. Since there would be no cycle, we add $e_2 = b - d$ and set $F = \{c - d, b - d\}$.

The next edge is $e_3 = b - c$. Now things get interesting. If we were to add $b - c$ to F, F would be $\{c - d, b - d, b - c\}$, which *does* have a cycle. Hence we *don't* add $b - c$ to F.

Adding $e_4 = a - d$ would mean that F would be $\{c - d, b - d, a - d\}$, which has no cycle, so we *do* add $e_4 = a - d$ to F. Now $F = \{c - d, b - d, a - d\}$.

Adding $e_5 = a - c$ would form a cycle, so we do *not* add $e_5 = a - c$ to F.

Last, adding $e_6 = a - b$ would also form a cycle. We do not add $e_6 = a - b$ to F.

Done. The final $F = \{c - d, b - d, a - d\}$, whose final cost is $3 + 5 + 7 = 15$. The amazing fact is that F not only is the edge set of a spanning tree, but also that that spanning tree has minimum cost among all spanning trees.

It is a theorem that a complete graph on n vertices has n^{n-2} spanning trees. Even if the number n of vertices is only 100, there are $100^{98} = (10^2)^{98} = 10^{196}$ spanning trees, that is, 1 followed by 196 zeros, or 10,000,000,000,000,000,000,000, 000,000,000,000,000,000,000,000,000,000,000,000,000,000,000,000,000,000,000,000, 000,000,000,000,000,000,000,000,000,000,000,000,000,000,000,000,000,000,000,000, 000,000,000,000,000,000,000,000,000,000,000,000,000,000,000,000,000,000,000,000, 000. Trust me, you don't want to find a minimum spanning tree by writing down all spanning trees, computing the cost of each, and then choosing the one of minimum cost. What's remarkable about the minimum spanning tree algorithm above is that it's fast and easy yet gives *exactly the same spanning tree* you would have found if you'd written down all 10^{196} spanning trees, computed the cost of each, and found the one of minimum cost.

There are two things to prove about the algorithm:

1. The set of edges one gets at the end defines a spanning tree, and
2. the total cost of that spanning tree is smallest among the costs of all spanning trees.

Theorem 9.15 *If $G = (V, E)$ is a connected graph, then if F^* is the set of edges at the end of running the algorithm and $G^* = (V, F^*)$, then G^* is a spanning tree.*

Proof The algorithm was careful never to add an edge if it would form a cycle. Hence at no time does the graph (V, F) contain a cycle, and, in particular, not at the end. Hence (V, F^*) contains no cycle. It clearly contains all vertices.

However, it's not obvious that (V, F^*) is connected. For all we know, F^* might be empty at the termination of the algorithm. In fact, if $G = (V, E)$ were itself not connected, then (V, F^*) wouldn't be connected either. This means that to prove that (V, F^*) is connected, the proof must use the fact that G is connected. (This is an important lesson you should not forget. If a lemma or theorem becomes false when a hypothesis is dropped, then that hypothesis must be used in the proof.)

Let us do a proof by contradiction. We will assume that $G^* = (V, F^*)$ is not connected, and derive a contradiction.

Suppose G^* is not connected. Now a connected graph is one with a path between every pair of vertices. Since G^* is not connected, there must be a pair of vertices, say, s and t, with no path in G^* between them. Since there is clearly always a (trivial) path between a vertex and itself, we know that $s \neq t$. Let S be the set of all vertices reachable by a path in G^* from s and let T be the set of all vertices reachable by a path in G^* from t. If there were a vertex, say, v, in both S and T, then there would be a path in G^* from s to v and from t to v, and hence (since paths are reversible) there would be a path in G^* from s to v to t, yet we know there is no path in G^* from s to t. It follows that no vertex is in both S and T. Are there any vertices in V which are neither in S nor T? There might be and there might not be.

Now G is promised to be a connected graph. It follows that G must have at least one edge with one endpoint in S and the other endpoint not in S. The reason is that if G had no edges "crossing the boundary" of S (i.e., with one endpoint in S and one not in S), then there could be no path from s to t, as such a path would have to have at least one edge crossing the boundary of S. This would contradict the fact that G is connected.

Let B be the set of edges of G which cross the boundary of S. By the previous paragraph, B has at least one edge. Let b be the cheapest edge of B. This edge b is the ith cheapest edge in the whole graph, for some integer i; that is, $b = e_i$. That edge e_i has two endpoints u, v, with one of them, say, u, in S, and the other, v, not in S.

Think about the point in time in running the algorithm when e_i was examined. At this point, the algorithm must have chosen not to take edge e_i and add it to F. The reason is simple. If $e_i = u - v$ had been included in F at the time, then e_i would have been in F^*, and since there was a path from s to u in G^*, there would also have been a path from s to v in G^*... but there couldn't have been, since v was not in S.

But now you have to wonder why the algorithm chose not to take edge e_i. The graph $G^* = (V, F^*)$ has no edge crossing the boundary of S, so earlier, at the time when e_i was considered, there must have been no edge crossing the boundary of S, either. So, clearly, adding e_i could not have created a cycle. This means the algorithm made a mistake by not taking e_i when it should have, which is not possible. The only assumption we made was that G^* was not connected. It follows that, in fact, G^* *was* connected. ∎

So now we know that at termination of the algorithm, $G^* = (V, F^*)$ is a spanning tree. But is it a *minimum* spanning tree? That is, does it have minimum cost among all spanning trees of G?

Theorem 9.16 *If $G = (V, E)$ is a connected graph, with, say, n vertices and m edges, and if F^* is the set of edges at the end of running the algorithm, then $G^* = (V, F^*)$ is a minimum spanning tree.*

Forgive me, but I am going to cheat a little bit in two ways. First, by assuming that all the m edges in G have different costs. This assumption is not really necessary, and it's easy to remove, but it makes the proof easier to understand. Second, I am going to take an example to illustrate the proof. Again, it's easy to turn the argument into a formal proof, but it's a little harder to understand, in my opinion.

Proof Sort the edges into order $e_1, e_2, e_3, ..., e_m$ such that $cost(e_1) < cost(e_2) < cost(e_3) < \cdots < cost(e_m)$. Let $G^* = (V, F^*)$ be the graph produced by running the minimum spanning tree algorithm. We will prove by contradiction that G^* is indeed a minimum spanning tree of G, so let us assume, for a contradiction, that G^* is not a minimum spanning tree. We know from Theorem 9.15 that it is a spanning tree so it must not have minimum cost. Let $G' = (V, F')$ be a minimum spanning tree in G, which exists by Theorem 9.13 or Theorem 9.15 (since if a spanning tree exists, so does a minimum spanning tree); its cost is less than the cost of G^*. Because all spanning trees on n-vertex graphs have $n - 1$ edges, both G^* and G' have $n - 1$ edges.

List the $n - 1$ edges of G^* in increasing order by cost. Similarly, list the $n - 1$ edges of G' in increasing order by cost. These two lists will agree for a while (maybe for 0 edges, maybe for more). Eventually, because the cost of G' is less than the cost of G^*, the lists must deviate. For example, the lists could both start with e_3, e_7, e_8, and then they could deviate. Maybe the list of G^* next has e_{10} while the list of G' next has e_{12}, or maybe it's reversed.

Suppose that the two lists both start identically, say, for example, with e_3, e_7, e_8, and then the list of G^* next has e_{i^*} while the list of G' next has $e_{i'}$, where $i^* \neq i'$. Since $i^* \neq i'$, there are two cases:

- Case 1: $i^* > i'$. Hence the cost of edge e_{i^*} is greater than the cost of edge $e_{i'}$ (as no two edges have the same cost). This is not possible. The reason is that G^* was generated by a greedy algorithm. Because $i' < i^*$, edge $e_{i'}$ was processed *before* e_{i^*} by the greedy algorithm. Since G' has no cycle, certainly the graph having edges $e_3, e_7, e_8, e_{i'}$ also has no cycle. But then the greedy algorithm should have taken edge $e_{i'}$ when it considered it (before getting to e_{i^*}), and it didn't, which is impossible.

- Case 2: $i^* < i'$. This case is more complicated. In this case, the greedy algorithm took an edge e_{i^*} of less cost than the edge $e_{i'}$ of the best tree.

  Take the edge set F' of the minimum spanning tree and consider adding to it edge e_{i^*} (which is definitely not already in it). Suppose e_{i^*} is the edge $u - v$. Now $G' = (V, F')$ is already a spanning tree, so, being connected, already has some path between u and v. Consider what would happen if we added edge $e_{i^*} = u - v$ to F'. It would have created a cycle which we'll call C. Since the greedy algorithm chose edges e_3, e_7, e_8, and e_{i^*}, we know that those four edges contain no cycle. Hence the cycle C must contain at least one edge e_j which is not among those four edges. All edges in F' other than e_3, e_7, e_8 have cost at least the cost of $e_{i'}$, which is greater than the cost of e_{i^*}; hence the cost of e_j exceeds the cost of e_{i^*}.

  Now here's the punchline: if you then *remove* that edge e_j, you get a tree again! That is, adding e_{i^*} to F' and then removing e_j gives a spanning tree. (Adding edge e_{i^*} to F' created a unique cycle C, and then removing e_j broke that cycle, leaving no cycles at all.) Furthermore, the edge e_{i^*} we added was cheaper than the edge e_j we removed, so the new spanning tree is cheaper than the one (V, F') with which we started. But that tree was already a minimum spanning tree, so this is a contradiction, as there can't be a cheaper spanning tree than the cheapest one.

Since we have reached a contradiction in both cases, the one assumption we made, that G^* is not a minimum spanning tree, must have been false. Hence G^* is a minimum spanning tree and we are done. ∎

This kind of argument is called an *exchange argument* as edge e_{i*} is exchanged for edge e_j.

9.6   Puzzle

A census taker knocked on the door of the home of a mathematician and asked for the ages of her children. She replied, "I have three children. The product of their ages is 36. The sum of their ages is the number on the house next door."

The census taker looked at the number on the house next door, thought for a moment, and said, "You did not give me enough information."

The mathematician also thought for a moment and said, "I apologize—you are right. My oldest child is at least a year older than all the others."

The census taker replied, "Thank you," and walked away.

How old were the children?

9.7   Exercises

Exercise 9.1 Prove that the sum of the degrees in a graph is always an even integer.

Exercise 9.2 Show that the relation that u and v are connected in a graph G is an equivalence relation.

Exercise 9.3 It is a theorem that a planar graph on n vertices has at most $3n - 3$ edges. Show that every planar graph has a vertex of degree at most five.

Exercise 9.4 * An *odd cycle* in a graph is a cycle with an odd number of vertices.

a) Show that the chromatic number of an odd cycle is 3.
b) Show that a graph can be two-colored if and only if it has no odd cycle.

Exercise 9.5 Prove that the complete graph K_n on n vertices has $n(n-1)/2$ edges.

Exercise 9.6 The *Petersen graph* appears in Fig. 9.14.

(a) Show that the Petersen graph cannot be colored with two colors.
(b) Show that the Petersen graph can be colored with three colors.

This proves that the chromatic number of the Petersen graph is 3.

Exercise 9.7 * Prove that the Petersen graph, shown in Fig. 9.14, is nonplanar.

Exercise 9.8 Consider the four-node complete graph K_4 shown in Fig. 9.13. This graph has 16 spanning trees. Draw and compute the weight of each. Determine

Fig. 9.14 The Petersen graph

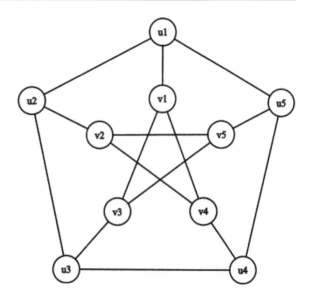

which spanning tree is cheapest (there may be a tie). Compare the cost of the cheapest spanning tree to that of the one found by the greedy algorithm.

Exercise 9.9

(a) Give an algorithm for MAXIMUM SPANNING TREE, the problem of finding a spanning tree the sum of whose edge weights is maximized (and prove it correct).
(b) Give an algorithm for MINIMUM PRODUCT SPANNING TREE. That is, give an algorithm for the following problem: given a connected graph with positive edge weights, find a spanning tree whose *product* (not sum) of edge weights is minimized.

Exercise 9.10 How many n-vertex graphs are there? Here we mean graphs as we've defined them. For example, the path $1 - 2 - 3$ on three vertices is a different graph from the path $2 - 1 - 3$.

Exercise 9.11 A *Hamiltonian cycle* in a graph is a cycle containing all the vertices of a graph. (Recall that a cycle cannot repeat vertices.) Known as HAMILTONIAN CYCLE, the problem of determining if a given graph has a Hamiltonian cycle appears to be very difficult.

In the TRAVELING SALESMAN problem, we are given an n-node complete graph G, $n \geq 3$, each of whose edges has a nonnegative length. Because the graph is complete and has at least three vertices, G has a Hamiltonian cycle. The goal is to find a Hamiltonian cycle of minimum length (length of the cycle being the sum of the lengths of the edges). To contrast HAMILTONIAN CYCLE with TRAVELING

SALESMAN, in the HAMILTONIAN CYCLE problem, the graph is not complete and
has no lengths.

Show that TRAVELING SALESMAN is at least as difficult as HAMILTONIAN
CYCLE, by showing that an algorithm for TRAVELING SALESMAN can be used
to solve HAMILTONIAN CYCLE.

Exercise 9.12 * Give an algorithm to determine if a graph is connected.

Exercise 9.13 * The problem 3-COLORABILITY of determining if an arbitrary
graph can be (vertex) colored with three or fewer colors seems very hard.

(a) Show that the problem 4-COLORABILITY of determining if a graph can be
    colored with four or fewer colors is at least as hard as 3-COLORABILITY. Do
    this by showing how an algorithm to solve 4-COLORABILITY could be used to
    solve 3-COLORABILITY.
(b) Show that the problem 5-COLORABILITY of determining if a graph can be
    colored with five or fewer colors is at least as hard as 3-COLORABILITY.
(c) Given a fixed positive integer k, k-COLORABILITY is the problem of determin-
    ing if a graph can be colored with k or fewer colors. Show that for all $k \geq 4$,
    k-COLORABILITY is at least as hard as 3-COLORABILITY.

Exercise 9.14 In the section on coloring planar graphs, we assumed that in a map,
each country consists of a *contiguous* region of the plane. In reality, this is not so;
for example, Alaska in the USA is not contiguous with the region defined by the 48
contiguous states.

Show that if countries may be defined by noncontiguous regions of the plane,
and if each country must be given only one color (e.g., Alaska must get the same
color as the region defined by the 48 contiguous states), then there are maps which
cannot be colored with four colors.

Exercise 9.15 * An *independent set S* in a graph is a set of vertices with no edges
between any pair of vertices in S. In a (vertex) coloring, the vertices of any one color
form an independent set. The *independence number* $\alpha(G)$ of a graph G is the size
of its largest independent set.

a) Prove that for any graph G on n vertices, $\chi(G) \geq n/\alpha(G)$.
b) The *complement* $\bar{G}$ of a graph G is obtained by switching edges and nonedges.
   That is, if u, v are distinct vertices, then $u - v$ is an edge of $\bar{G}$ if and only if $u - v$
   is not an edge of G.
       Show that $\chi(\bar{G}) \geq \alpha(G)$.
c) Show that

$$\chi(G) + \chi(\bar{G}) \geq \sqrt{n}.$$

d) If you know calculus, show that

$$\chi(G) + \chi(\bar{G}) \geq 2\sqrt{n}.$$

Exercise 9.16 * A *matching* in a graph is a set of edges, no two sharing an endpoint. In an edge coloring of a graph, the edges of one color always form a matching. The *matching number* $v(G)$ (where v is yet another Greek letter, called *nu*, and pronounced "new") is the size of the largest matching in a graph:

a) Let G be a graph with m edges. Show that $\chi'(G) \geq m/v(G)$.
b) Show that $v(K_n) = n/2$ if n is even and $v(K_n) = (n - 1)/2$ if n is odd.
c) Suppose n is even.
   1. Show that $\chi'(K_n) \geq n - 1$.
   2. Show that $\chi'(K_n) \leq n - 1$. This is very difficult.
   3. Conclude that $\chi'(K_n) = n - 1$ if n is even.
d) Suppose n is odd.
   1. Show that $\chi'(K_n) \geq n$.
   2. Show that $\chi'(K_n) \leq \chi'(K_{n+1})$. Use this fact to show that $\chi'(K_n) \leq n$.
   3. Conclude that $\chi'(K_n) = n$ if n is odd.

Exercise 9.17 A *triangle-free graph* is one containing no K_3. Show that if n is even, there is a triangle-free graph on n vertices with $n^2/4$ edges.

Exercise 9.18 Clearly if a graph contains a clique of size r, then the chromatic number is at least r. But is this the only way a graph can have large chromatic number?

A *triangle-free graph* contains no K_3. It is easy to construct a triangle-free graph of chromatic number 3, say, the cycle C_5 of length five, but can you construct a triangle-free graph of chromatic number 4? The Grötzsch graph, which is triangle-free, is shown in Fig. 9.15. Show it cannot be colored with three colors.

Interestingly, it is known that for every positive integer c, there is a triangle-free graph of chromatic number c. One way to construct such graphs is to repeat the construction that was used to convert the 5-cycle C_5 (on u_1, u_2, u_3, u_4, u_5) into the 11-vertex Grötzsch graph.

Exercise 9.19 * In a graph $G = (V, E)$, a *happy coloring* is a coloring of the vertex set using two colors M and F so that for each vertex v, at least half of its neighbors are of the opposite color from v. (Note that this is not a coloring in the sense of vertex coloring described earlier.) For example, if a vertex with six neighbors is colored M, at least three of its neighbors must be colored F.

Show that every graph has a happy coloring.

Exercise 9.20 * Show every six-vertex graph contains either a clique of size 3 or an independent set of size 3.

Interestingly, this result can be generalized to prove this theorem.

Fig. 9.15 The Grötzsch
graph. Vertices
u_1, u_2, u_3, u_4, u_5, in order,
define a five-cycle. Each v_i is
adjacent to all of u_i's
neighbors in the cycle and to
one additional vertex, w

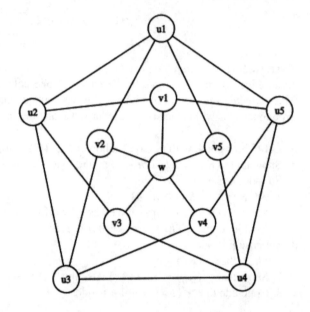

Theorem 9.17 *For any positive integers r, s, there is a (finite) number $R(r, s)$ such that any graph with at least $R(r, s)$ vertices contains either a clique of size r or an independent set of size s.*

It is far from obvious that such an $R(r, s)$ exists. This theorem is part of *Ramsey Theory*.

Exercise 9.21 One might question whether a stronger theorem than Vizing's Theorem holds, at least if G is a multigraph and its multiplicity $\mu(G)$ is large. For example, here is a conjecture.

Conjecture 9.3 For any multigraph G of maximum degree $\Delta(G)$, having multiplicity $\mu(G)$ at least 10, $\chi'(G) \leq \Delta(G) + (1/2)\mu(G)$.

Show that this conjecture is false.

Exercise 9.22 A *vertex cover* S in a graph is a subset of vertices containing at least one endpoint of every edge. VERTEX COVER is the problem of finding a smallest vertex cover in a graph, a problem that appears to be very difficult. The size of the smallest vertex cover in G is denoted $\tau(G)$, "tau" (rhyming with "ow!") being yet another Greek letter. Because finding a smallest vertex cover is so hard, we will instead find a vertex cover of size at most $2\tau(G)$, that is, at most twice the size of the smallest vertex cover.

A *matching* in a graph G is a collection of edges having no vertices in common, that is, t edges on $2t$ distinct endpoints. A *greedy matching* M is found by (1) taking any one edge e_1; (2) taking a second edge e_2 which avoids both endpoints of e_1; (3) taking a third edge which avoids the four endpoints of e_1 and e_2; and repeating, as long as possible. Let h be the number of edges in the greedy matching (it doesn't matter which greedy matching you choose, but it's important when you build the greedy matching that you keep adding edges as long as possible).

(a) Show that any vertex cover must contain at least one vertex from each edge in M. (What if it missed both endpoints of an edge in M?) Hence the size $\tau(G)$ of the smallest vertex cover of G must be at least the size $|M|$ of M (which is the number of edges in the matching).
(b) Let T be the set of *all* endpoints of all edges in the greedy matching M, a set of size $2|M|$ (because each edge has two endpoints). Prove that T is a vertex cover. (Ask yourself how it could fail to be a vertex cover.)
(c) Show that the size $|T|$ of the vertex cover found in (b) is at most twice the size of the smallest vertex cover. This is known as a *two-approximation algorithm* since it generates a solution of size at most twice the optimal size.

Exercise 9.23 See Exercise 9.22 for the definition of a vertex cover and of the VERTEX COVER problem.

Let $G = (V, E)$ be a graph. The *complement* of a set $S \subseteq V$ of vertices is the set $V - S$ of vertices not in S.

Let $S \subseteq V$.

(a) Show that if S is independent (i.e., there are no edges between vertices in S), then its complement $V - S$ is a vertex cover.
(b) Show that if a subset T is a vertex cover, then its complement $V - T$ is an independent set.
(c) INDEPENDENT SET is the problem of finding a largest independent set in a graph. Show how any algorithm for INDEPENDENT SET can be used to solve VERTEX COVER and *vice versa*.
(d) In light of the perceived difficulty of VERTEX COVER, argue that INDEPENDENT SET must be difficult to solve as well.

9.7.1 Hints

9.4 Do this by giving an algorithm which attempts to color a graph with two colors, and show that if the algorithm fails, one can identify an odd cycle in the graph.
9.7 Find a subdivision of $K_{3,3}$ in the Petersen graph.
9.12 Start with all vertices unlabeled. Pick any vertex, say, vertex s, and give it label 0.

Give label 1 to all neighbors of vertex s (if there are any).
Give label 2 to all unlabeled neighbors of all vertices having label 1 (if there are any).

Give label 3 to all unlabeled neighbors of all vertices having label 2 (if there are any).

Give label 4 to all unlabeled neighbors of all vertices having label 3 (if there are any).

Repeat.

Then what?

9.13 Add one or more new vertices adjacent to all original vertices and to each other.

9.15 (c). Think about two cases: $\alpha(G)$ is small and $\alpha(G)$ is large.

9.16 (c). Arrange vertices 1, 2, 3, ..., $n-1$ of K_n as the vertices of a regular $(n-1)$-gon C, in order, and place the last vertex, vertex n, in the center. Say two edges (drawn as line segments) are *perpendicular* if the (infinite) line containing the first line segment is perpendicular to the (infinite) line containing the second line segment. Now color with color i, $i = 1, 2, 3, \ldots, n - 1$, the "radial" edge e_i from the center to vertex i, as well as all edges, between vertices of C, which are perpendicular to edge e_i. Argue that this is a valid edge coloring. (This hint is from [1].)

9.19 Start by assigning every vertex an arbitrary color. If a vertex is unhappy, change its color. Prove this process must terminate and show that when it terminates, the graph is happily colored.

9.20 Suppose, for a contradiction, that G is a six-vertex graph containing neither a clique nor an independent set of size 3. First argue that the maximum degree in G must be two. Then argue that even if the maximum degree is two, there must be a clique or independent set of size 3.

Probability

10

10.1 Examples

Suppose you have a bag full of n identical balls except that they are labeled 1, 2, 3, ..., n. To be concrete, let's suppose there are $n = 5$ balls (but there is nothing special about five). You stick your hand in and randomly pick one of the balls without looking. What's the chance you choose ball number 2? There are five balls, so there are five possibilities: that you choose ball number 1, ball number 2, ball number 3, ball number 4, or ball number 5. All of the possibilities are equally likely. Probabilities always add to 1. Since the five outcomes should all have the same probability, and they should add to 1, each probability should be 1/5. The chance of choosing ball 2 is 1/5.

What is the probability of choosing a ball with an even number? The even numbers in the set $\{1, 2, 3, 4, 5\}$ are 2 and 4, so the probability of choosing an even-numbered ball is 2/5.

What is the probability of choosing a prime-numbered ball? The primes in $\{1, 2, 3, 4, 5\}$ are 2, 3, 5, so the probability of choosing a prime-numbered ball is 3/5.

An *event* is a set of possible outcomes of an experiment. In our situation, with identical (but labeled) balls, an event is a subset S of $\{1, 2, 3, ..., n\}$ and the probability of an event is the size of S, divided by n. For example, in the case in which we want to pick a prime-numbered ball, $S = \{2, 3, 5\}$ and the probability is 3/5.

Here's another example. Suppose that every year has 365 days (i.e., that there are no leap years) and that children are born on uniformly random days of the year. What's the chance that a child was born in December? There are 31 days in December, out of 365 in a year, so the chance that a random child was born in December is 31/365.

© The Author(s), under exclusive license to Springer Nature Switzerland AG 2023
H. Karloff, *Mathematical Thinking*, Compact Textbooks in Mathematics,
https://doi.org/10.1007/978-3-031-33203-6_10

Let's call the objects we're choosing *points*. You can imagine that there are n points labeled $1, 2, 3, ..., n$. In the birthday example, there are 365 points, since there are 365 possible outcomes, and they are all equally likely.

By the way, I keep mentioning, "They are all equally likely" because it is a crucial assumption. When there are two equally likely outcomes, and one or the other must occur, the probability of each outcome is $1/2$. However, there may be two outcomes of differing probabilities. For example, you board a flight to Hawaii. Either the plane crashes or it does not. The possibility that the plane crashes is extremely unlikely, and so the two probabilities are not the same. The chance the plane arrives safely is almost 1, and the chance that the plane crashes is incredibly small, almost 0.

Here's a hard question. If you take *two* random people, what's the chance that they have the same birthday? Let's write down the outcome of the experiment as a pair (i, j) where i is the day of the year on which the first person was born and j is the day of the year on which the second was born. It is important to realize that outcomes $(3, 5)$ and $(5, 3)$ are different. There are 365 possible i's and 365 possible j's and hence 365^2 possible equally likely points. The set of points in which the two people have the same birthday is $\{(1, 1), (2, 2), (3, 3), ..., (365, 365)\}$, of size 365. Altogether, there are $n = 365^2$ points. If you wish, think of $n = 365^2$ balls labeled, not $1, 2, 3, ..., n$, but $(1, 1), (1, 2), ..., (1, 365); (2, 1), (2, 2), ..., (2, 365); (3, 1), (3, 2), ..., (3, 365); ...; (365, 1), (365, 2), ..., (365, 365)$. Hence, the probability that two people have the same birthday is $365/365^2 = 1/365$.

Wait, there's an easier way to see the same thing. Think of choosing random birthdays for two people *one at a time*. We choose the random birthday i of the first person, and then, and only then, we choose the random birthday j of the second person. Whatever the value of i, the chance that j equals i is $1/365$, since there is only one possible value of j that equals i. Since this is true for all i, the probability that $i = j$ is $1/365$.

10.2 Independence

The probability of an event E is denoted $P[E]$. If E_1 and E_2 are events, the event that both E_1 and E_2 occur is denoted $E_1 \wedge E_2$, where "$\wedge$" means "and."

The standard definition of independence is that two events E_1 and E_2 are independent if the probability $P[E_1 \wedge E_2]$ that both E_1 and E_2 occur is the probability that E_1 occurs multiplied by the probability that E_2 occurs. That is, formally, E_1 and E_2 are *independent* if

$$P[E_1 \wedge E_2] = P[E_1] \cdot P[E_2].$$

This is the correct, standard definition, but one misses the point of independence if it is unmotivated.

I will motivate this definition by example. Let's first consider a bizarre experiment. You roll a normal six-sided die D_1 on a table, and then you take a second die D_2 and place it on the table in the same orientation as D_1. Now one looks at the numbers atop the two dice. Since we're doing an unnatural experiment, there are only six possible outcomes: $(1, 1)$ (meaning D_1 and D_2 are showing 1's), $(2, 2)$ (meaning both are showing 2's), $(3, 3)$, $(4, 4)$, $(5, 5)$, and $(6, 6)$. Let event E_i^1 be the event that the first die shows i and let E_j^2 be the event that the second die shows j. Then for $i = 1, 2, 3, 4, 5, 6$, $P[E_i^1] = P[E_i^2] = 1/6$. Furthermore, and this is the interesting point, $P[E_i^1 \wedge E_i^2] = 1/6$ also. Note that $P[E_3^1 \wedge E_5^2] = 0$, that is, the probability that the first die shows a 3 and the second die shows a 5 is 0. Note also that in this bizarre experiment, the two outcomes are intimately connected: as soon as you know how D_1 has landed, you know exactly how D_2 has landed. This is the opposite of independence! If the events that the first die landed with a 3 and the second landed with a 5 were independent, then the probability that both events happened would be $(1/6)(1/6) = 1/36$ and not 0. Furthermore, the chance that both dice landed with 1's would be $1/36$ instead of $1/6$.

Now let's think about rolling two dice in the usual way. Assuming the two rolls are unrelated, the chance of getting any fixed pair (i, j) of the 36 possible pairs $(1, 1)$, $(1, 2)$,..., $(6, 6)$ is $1/36$, which is exactly the product of the probability of getting i on the first die and the probability of getting j on the second die. Before rolling any dice, the chance that the second die D_2 shows a 5 is $1/6$. Now if I tell you that the first die rolled a 3, knowing that information, *still*, the chance the second die shows a 5 is $1/6$. This was definitely not the case in the bizarre experiment we did earlier—knowing how D_1 landed told us the outcome of D_2 with certainty. Two events are independent if knowing the outcome of the first event doesn't change the probability of the second event. This certainly happens if the events are unrelated, that is, if neither influences the other.

We generalize independence to more than two events by saying that events $E_1, E_2, ..., E_n$ are independent if knowing the outcomes of all of the events except for E_i tells us nothing about the outcome of E_i (for each i). For example, if $n = 3$, knowing the outcomes of E_1 and E_2 tells us nothing about E_3; knowing the outcomes of E_1 and E_3 tells us nothing about E_2; and knowing the outcomes of E_2 and E_3 tells us nothing about E_1. If events are independent, then the probabilities multiply. For example, the probability that E_1, E_2, and E_3 all occur is the product of $P[E_1]$, $P[E_2]$, and $P[E_3]$.

10.3   Birthday Parties

10.3.1 Hitting a Specific Date

If you have n random people at a party, what's the chance that someone at the party was born on January 1st (ignoring leap years)? Here's a follow-up question: how large must n be so that it's likely that someone was born on January 1st, where "likely" means, "has probability more than $1/2$?"

It's easier to compute the probability that *no one* was born on January 1st. Let E_i be the event that person i was *not* born on January 1st. If n is 1, the chance that that person was not born on January 1st is $364/365$. Now let n be 2. Knowing whether person 1 was or was not born on January 1st tells us nothing about whether person 2 was born on January 1st, so the events E_1 and E_2 are independent. This means we get to multiply probabilities: $P[E_1 \wedge E_2] = P[E_1] \cdot P[E_2] = \frac{364}{365} \frac{364}{365} = \left(\frac{364}{365}\right)^2$.

In general, the probability that all n people were *not* born on January 1st is $\left(\frac{364}{365}\right)^n$, since the set of all n events is independent.

The *complement* $\bar{E}$ of an event E is the event made up precisely of the points which are not in E. Since the total probability is always 1, $P[E] + P[\bar{E}] = 1$ always and hence $P[\bar{E}] = 1 - P[E]$ for any event E.

In the current example, E is the event that no partygoer has a birthday on January 1st. The complement $\bar{E}$ is the event that at least one partygoer has a birthday on January 1st. (Make sure you understand why the complement is not that *all* partygoers have birthdays on January 1st.) Hence the chance that some party attendee has a birthday on January 1st is $1 - (364/365)^n$. Hence the smallest n so that it's likely that some partygoer has a birthday on January 1 is the least n such that $(364/365)^n < 1/2$. Some playing on a calculator or taking logarithms of both sides tells us that $n = 253$.

It is surprising to me, even now, that the answer isn't half of 365, rounded up, or 183.

10.3.2 Getting Two Identical Birthdays

We just saw that one needs to have 253 people at a party in order for it to be likely that someone has been born on January 1st. Instead, how many people do we need at the party for it to be likely that two people have the same birthday (which need not be January 1st)? Take a guess. How big a party do you need?

The answer is surprisingly few: only 23. The next time you're at a party (or in a classroom) with 23 or more people there, ask around, and don't be surprised if two people have the same birthday.

Here's the proof. Once again, we will compute the complementary probability, which is the probability that no two people have the same birthday, and we'll look for the smallest n such that this probability is less than $1/2$.

Let the birthday of partygoer i be B_i, which is in the set $\{1, 2, 3, ..., 365\}$. If in the sequence $< B_1, B_2, B_3, ..., B_n >$ all the B's are different, then obviously in any prefix $< B_1, B_2, B_3, ..., B_i >$ (if $i \leq n$), all the birthdays are different. This provides an idea as to how to compute the probability. Start with a probability estimate of 1.

1. Choose B_1 randomly, and multiply the probability estimate by the probability that all of B_1 are different (which is clearly 1, because there's only one birthday);
2. Choose B_2 randomly, and multiply the probability estimate by the probability that B_1 and B_2 are different (which is 364/365);

3. Assuming that B_1 and B_2 are different, choose B_3 randomly, and multiply the probability estimate by the probability that B_1, B_2, and B_3 are all different (which is $363/365$);

4. Assuming that B_1, B_2, and B_3 are different, choose B_4 randomly, and multiply the probability estimate by the probability that B_1, B_2, B_3, and B_4 are all different (which is $362/365$);

5. Assuming that B_1, B_2, B_3, and B_4 are different, choose B_5 randomly, and multiply the probability estimate by the probability that B_1, B_2, B_3, B_4, and B_5 are all different (which is $361/365$);

6. etc.

Notice that the factor in step n is $(366 - n)/365$.

Continuing in this way, for general n, the chance that B_1, B_2, B_3, ..., B_n are all different is

$$(1) \cdot (364/365) \cdot (363/365) \cdot (362/365) \cdots ((366 - n)/365).$$

This means that we want the smallest n such that this quantity is less than $1/2$.

At this point it's time to pull out the trusty calculator or whip up some Python code. Here's the Python code:

```
product = 1
n = 0
while product >= 0.5:
    n = n + 1
    product *= (366 - n)/365
    print("For n=", n, "product=", product)
```

If you run the code yourself, you will see that product drops below 0.5 for the first time at $n = 23$.

10.4 Gambling

A *lottery* is a random game, often run by the government, with a specified payoff. Here is a concrete example. You pay \$1 to select an integer between 1 and 1000. Later, the lottery operator picks an integer in the same range, and if it agrees with your number, you get \$500. Having wagered only \$1, you might think this sounds like a quick way to get rich. Is it?

Well, let's figure it out. There are 1000 possible winning numbers. Your chance of picking the right one is $1/1000$. If you were to play the game many times, you'd win roughly $1/1000$ of the games. This means that on average, your expected winnings would be $(1/1000) \cdot \$500$, or \$0.50. You paid \$1 for the ticket, so your *expected profit* would be $\$0.50 - \$1$, or $-\$0.50$. On average when you play the game, you lose 50 cents. This is a very bad game which you should never play. Generally you should never play a game in which your expected profit is negative. Save your money for games with positive expected profit.

By the way, state-run lotteries are almost always just complicated versions of this game. You should (almost) never buy lottery tickets.

However, there are rare times when it can be profitable to buy lottery tickets. Specifically, there are rare lotteries, for example, some Powerball lotteries in the USA, with a jackpot so high that it can be profitable to buy a ticket. Whether it is profitable or not depends on how high the jackpot is, how many people buy tickets (because you might have to share the winnings), how much you'd have to pay in taxes, when you would get your winnings, how much the money is worth to you ($100 million is worth less in practice than a hundred times $1 million), etc. The calculation of expected profit is complicated. In general, except possibly for state lotteries with enormous jackpots, it's a bad idea to buy lottery tickets.

That doesn't mean it's always bad to gamble. It's good to gamble when the odds are in your favor.

Many years ago, when I started college, there was a "casino night" during orientation week during which people gambled with play money. My dorm pooled its fake money, bid on, and won a stereo system. This stereo was dorm-owned and no one knew what to do with it... so it sat under someone's bed all year. The end of the year came and still no one knew what to do with it, so it was decided to raffle it off. In most raffles, the cost of the prize is dwarfed by the value of tickets sold. In this raffle, it appeared that the total value of tickets sold might be *less* than the value of the prize (which I estimated to be $100), which might make it a raffle with positive expected profit.

I decided I would try to be the last to buy tickets, which would sell for $0.25 each. If n tickets had been sold before I bought any and I bought x, my chance of winning would be $x/(n + x)$. Given the stereo was worth $100, my expected winnings would be $(\frac{x}{n+x})100$. The cost of the tickets would be $0.25x$. Hence my expected profit would be

$$p(x) = \left(\frac{x}{n + x} \right) 100 - 0.25x.$$

The question then was, how many tickets should I buy? In other words, which x, as a function of n, maximizes the expected profit?

It is an exercise, requiring calculus, to find the x, as a function of n, which maximizes $p(x)$.

The day of the raffle I stood alongside the ticket seller, asking him repeatedly how many tickets he'd sold. At the very end, he told me he'd sold 180 tickets. I bought approximately the optimal number of tickets, and to my glee, I won. (See Exercise 10.5.)

You too can use probability for fun and profit.

10.5   Let's Make a Deal

I'm going to give you now a puzzle, in fact, probably my all-time favorite puzzle. This puzzle is modeled on an old TV game show called "Let's Make a Deal." Here's how (part of) the game show ran. For a long time, the host of the game show was named Monte Hall. Monte would bring a contestant out of the audience onto the stage and then announce, "Welcome, Maria [or Susan or Charles], to Let's Make a Deal! You see behind me three doors: door number 1, door number 2, and door number 3. Behind one of the doors sits a car. Behind the other two sit goats.[1] You get to keep whatever's behind the door you pick. Which door would you like?"

The audience goes crazy. One third are yelling "Door number 1!", one third, "Door number 2!", and one third, "Door number 3!". Of course the audience had no idea where the car was.

The contestant then picks a door, say, door number 1. Now Monte picks one of the two remaining doors, in this case door 2 or door 3, and says, "Here's what you passed up by not choosing door 2" or "Here's what you passed up by not choosing door number 3." Let's say Monte chooses door number 2. Door number 2 is opened and behind it sits a goat. (There was never a car behind the selected door.)

Now Monte says, "You can stick with your original choice of door 1 or you can switch to door number 3. What would you like to do?" Again the audience goes wild. Half are shouting, "Stick!", and the other half are shouting, "Switch!". The contestant decides to stick or switch; the chosen door is opened; behind it sits a car or a goat; and the contestant goes home either richer or with a new pet.

Now the question: what should the contestant do? Should the contestant stick with his or her original choice, switch to the other door, or does it not matter?

I am embarrassed to admit when I heard the puzzle originally, I got it wrong. Figure out what the contestant should do! Please, this is too good a puzzle for you not to think about it. Don't read ahead until you're ready for the answer!

Let's analyze the game. Assume that the car is put behind a random door and that goats are worth nothing relative to a car, though I imagine that having a goat could save a homeowner the cost of mowing the lawn.

There are two strategies for the contestant, when given the choice: (1) stick with the original choice, or (2) switch to the other door.

In strategy 1, you win if, and only if, the car is behind your original choice of door; otherwise you don't. All of that hullabaloo about switching is irrelevant. Since the car is behind a random door, the contestant wins with probability 1/3.

In strategy 2, you win if, and only if, the car is *not* behind the original door. Let's check this. Let's say the contestant originally picks door number 1, as above. The car is equally likely to be behind door number 1, door number 2, or door number 3. If the car is behind the chosen door, door number 1, then Monte has the option

---

[1] While the booby prizes may not always have been goats, let's just pretend that they were.

to show what's behind door number 2 or door number 3. Whichever one he picks, when the contestant switches, he or she loses.

But what if the car is behind a different door, door number 2 or door number 3?

If the car is behind door number 2, Monte *must* reveal what's behind door number 3. Now the contestant switches to door number 2, and wins.

Similarly, if the car is behind door number 3, Monte reveals what's behind door number 2. Now the contest switches to door number 3 and wins.

So in the "switching" case, the contestant wins two times out of three, for a probability of 2/3 of winning.

The conclusion is that the contestant should always switch. I got the answer wrong when I first heard this puzzle. I flippantly thought, "It doesn't matter," because I didn't work out the solution carefully. The moral of the story is think carefully about a problem before answering.

10.6   The (Magical) Probabilistic Method

Sometimes probability is used to prove a theorem that has nothing to do with probability. These are real proofs—there is no doubt about their correctness. Often probability is used to prove that a certain object *exists*, without actually finding it. If you've never seen such a proof, it can seem kind of magical. How could one prove something exists without finding it?

I imagine a good analogy is with the *pigeonhole principle*: if you put $n + 1$ pigeons[2] into n pigeonholes, some pigeonhole gets two or more pigeons. This is certainly true, but it gives you no idea how to find the pigeonhole having multiple pigeons.

In Chap. 9 on graph theory, in the exercises, you learned about *Ramsey numbers* $r(k, l)$. Ramsey number $r(k, l)$ is the smallest integer n such that every graph having n (or more) vertices contains either a clique of size k or an independent set of size l. Alternatively, $r(k, l)$ is the smallest positive integer n such that if the edges of K_n are colored red or blue, there is either a red K_r or a blue K_l. It is far from obvious that there even *is* a number with this property.

To state an upper bound on $r(k, l)$, I have to remind you of $n!$ (read "n factorial") and the binomial coefficient $\binom{n}{k}$. First, $n!$ is just the product of the first n positive integers: $n! = n \cdot (n - 1) \cdot (n - 2) \cdots 3 \cdot 2 \cdot 1$. For example, $3! = 3 \cdot 2 \cdot 1 = 6$. This is just the number of ways of writing down n distinct symbols in all possible orderings (known as *permutations*): there are n choices for what comes first, $n - 1$ choices for what comes next, $n - 2$ choices for what appears in the third position, and so on. If you take $n = 3$ and symbols a, b, and c, there are $3! = 6$ permutations: abc, acb, bac, bca, cab, and cba.

---

[2] Or anything else for that matter—no pigeons were harmed in writing this chapter.

Definition 10.1 The *binomial coefficient* $\binom{n}{k}$ is the number of k-element subsets of an n-element set.

For example, there are 10 2-element subsets of $\{1, 2, 3, 4, 5\}$, namely, $\{1, 2\}$, $\{1, 3\}$, $\{1, 4\}$, $\{1, 5\}$, $\{2, 3\}$, $\{2, 4\}$, $\{2, 5\}$, $\{3, 4\}$, $\{3, 5\}$, and $\{4, 5\}$, and hence $\binom{5}{2} = 10$. To calculate $\binom{n}{k}$, we first write down any k *distinct* elements of $\{1, 2, 3, ..., n\}$. There are n choices for the first element, $n-1$ for the second element, $n-2$ for the third, ..., and finally $n-k+1$ for the kth element. The product $n \cdot (n-1) \cdot (n-2) \cdots (n-k+1)$ equals $n!/(n-k)!$. But now we see that each of the k-element subsets of $\{1, 2, ..., n\}$ has appeared multiple times. How many times has each subset appeared? Once for each permutation of the k integers chosen, and there are $k!$ permutations. It follows that we have to divide $n!/(n-k)!$ by the number of times, $k!$, that each subset appears. It follows that $\binom{n}{k} = \frac{n!}{(n-k)!k!}$. For example, we know that $\binom{5}{2} = 10$. Fortunately, $5!/(3!2!) = 120/(6 \cdot 2) = 120/12 = 10$.

Note that, since $n \cdot (n-1) \cdot (n-2) \cdots (n-k+1) \le n^k$, we clearly have $\binom{n}{k} \le n^k/k!$. We also have $\binom{n}{k} \le 2^n$.

Now let's get back to the Ramsey numbers $r(k, l)$. Here is an upper bound on $r(k, l)$.

Theorem 10.1

$$r(k, l) \le \binom{k+l-2}{k-1} = \binom{k+l-2}{l-1} \le 2^{k+l-2}.$$

Hence

$$r(k, k) \le \binom{2k-2}{k-1} \le 2^{2k-2}.$$

This theorem is not hard to prove by induction but I will not include a proof. Notice that it states that $r(3, 3) \le \binom{6-2}{2} = \binom{4}{2} = 6$. In fact, in an exercise of Chap. 9, you proved that $r(3, 3) \le 6$. The question is, could $r(3, 3)$ be smaller than 6? Could it be 5, for example? It could not. The five-cycle C_5 has no clique of size 3 and no independent set of size 3. Hence $r(3, 3) > 5$. It follows that $r(3, 3) = 6$.

How can we prove a lower bound on $r(k, k)$ for $k > 3$? It turns out that the best way to do so is to use the "probabilistic method." This seemingly magical method allows you to prove that an object exists without finding it.

Before we get started, let me introduce Hungarian mathematician Paul Erdős, inventor of the probabilistic method. Paul Erdős wrote approximately 1500 mathematical papers, making him one of the most prolific mathematicians of all time. He lived an itinerant lifestyle with the sole goal of writing joint research papers. He worked with 509 collaborators in his lifetime. He wrote so many papers that his work spawned the concept of *Erdős number*, which is the minimum number of "hops" of joint research papers one has follow to get from Erdős to a given person. Specifically, Erdős has Erdős number 0; any person who coauthored a paper with

Erdős has Erdős number 1; any other person who coauthored a paper with someone of Erdős number 1 has Erdős number 2; any other person who coauthored a paper with someone of Erdős number 2 has Erdős number 3; and so on. There are over 11,000 people having Erdős number 2 and I am pleased to be one of them. In fact, I had the pleasure of sitting next to Erdős one night at a dinner near the University of Chicago.

Speaking of the University of Chicago, there is an old apocryphal anecdote told about one of Erdős's visits to the University of Chicago. Erdős liked to walk while thinking about math problems. During a visit to the University of Chicago as an old man, Erdős went for a walk one night to think hard. His walk took him off campus into the surrounding neighborhood. Oblivious as usual to his surroundings, Erdős just walked, and thought.

A police car stopped, the officer concerned about the meandering old man. "What are you doing here at this time of night?", the officer asked.

"Thinking," replied Erdős.

"Get in the car," replied the officer.

The officer took Erdős to the police station. At the station, a more senior officer interrogated Erdős. "What were you doing wandering around alone so late at night?", he asked, possibly fearing that Erdős had Alzheimer's.

"Thinking," again replied Erdős.

This officer tried to go deeper than the first one. "About what?", he asked.

"A conjecture," replied Erdős.

"What conjecture?", asked the officer.

Erdős replied, "It doesn't matter. It's false."

I guess he disproved it in the police car.

Erdős is famous for many of his quotes, one of his most famous being "A mathematician is a device for turning coffee into theorems".[3] Erdős popularized the concept of "The Book," which contains the shortest and most beautiful proofs. When Erdős saw a proof he particularly liked, he exclaimed, "That's straight from The Book." Erdős explained, "You don't have to believe in God, but you should believe in The Book."

Several books have been written about Erdős. You can read more about Erdős on the Web.

Let's get back to the Ramsey numbers $r(k, l)$. Computing the Ramsey numbers $r(k, l)$ appears to be an extremely difficult problem. After all, computing, say, $r(5, 5)$ exactly would mean finding an n such that:

1. Every n-node graph contains either a clique of size 5 or an independent set of size 5, and, furthermore,
2. this is not true for $n - 1$, that is, there is some $(n - 1)$-node graph which contains neither a clique of size 5 nor an independent set of size 5.

---

[3] In which case I'm doomed, since I don't drink coffee.

This sounds very hard. In fact, the great Erdős himself had this to say about the difficulty of the problem: "Suppose aliens invade the earth and threaten to obliterate it in a year's time unless human beings can find the Ramsey number for red five and blue five. We could marshal the world's best minds and fastest computers, and within a year we could probably calculate the value. If the aliens demanded the Ramsey number for red six and blue six, however, we would have no choice but to launch a preemptive attack" [4].

Since computing the Ramsey numbers exactly is so difficult, let's focus on getting good lower and upper bounds on $r(k, l)$. We already saw an upper bound. Erdős gave a beautiful simple proof that for $k \geq 4$, $r(k, k) > 2^{k/2}$. To do so, he introduced the *probabilistic method*. This seemingly magical method is often used to prove objects exist without constructing them. Of course, one can always find the desired object, once it is known to exist, by trying all the possibilities, but doing that's usually so slow it's useless.

We will need the simple *union bound* from basic set theory. You remember that the size of the union of A and B is at most the size of A plus the size of B. More generally, we have this theorem.

Theorem 10.2 *The size of the union of a finite collection of finite sets is at most the sum of the sizes of the sets.*

Now we're ready to prove the Ramsey number lower bound. Here $\lfloor x \rfloor$, read "the floor of x," is x itself if x is an integer, and the integer "just under" x otherwise.

Theorem 10.3 *Fix any $k \geq 4$ and let $n = \lfloor 2^{k/2} \rfloor \geq k$. There is an n-node graph having neither a k-node clique nor a k-node independent set.*

Corollary 10.1 *For $k \geq 4$, $r(k, k) > 2^{k/2}$.*

Proof The theorem shows that $r(k, k) > \lfloor 2^{k/2} \rfloor$. If $2^{k/2}$ is an integer, we have $r(k, k) > \lfloor 2^{k/2} \rfloor = 2^{k/2}$ and we are done. If not, we have $r(k, k) > \lfloor 2^{k/2} \rfloor$. Since $r(k, k)$ is an integer and $2^{k/2}$ is not, we again have $r(k, k) > 2^{k/2}$, and we are done. ∎

Proof of Theorem 10.3 Choose any positive integer $k \geq 4$. Let $n = \lfloor 2^{k/2} \rfloor$.

We will show (1) that the fraction p of all n-node graphs containing a k-node clique is at most $1/6$ and (2) that the fraction q of all n-node graphs containing a k-node independent set is also at most $1/6$. Because $1/6 + 1/6 = 1/3 < 1$, there is a positive probability of at least $1 - p - q \geq 2/3$ that a random n-node graph contains no clique of size k and no independent set of size k, and hence the desired graph must exist.

By the way, it is not hard to see that $p = q$. The reason is that for each random graph we could have chosen, we could equally well have chosen its complement. Under complementation (switching edges and nonedges), cliques become independent sets and independent sets become cliques.

We will prove $p \le 1/6$ by exhibiting a very simple upper bound on p and then showing that the upper bound is itself at most $1/6$. For example, suppose we want to count the number of five-node graphs containing a 3-clique. The 3-clique could be $\{1, 2, 3\}$, or $\{1, 2, 4\}$, or $\{1, 2, 5\}$, or $\{1, 3, 4\}$, or $\{1, 3, 5\}$, or $\{1, 4, 5\}$, or $\{2, 3, 4\}$, or $\{2, 3, 5\}$, or $\{2, 4, 5\}$, or $\{3, 4, 5\}$. In total there are $\binom{5}{3} = 10$ possible 3-cliques. The number of five-node graphs containing a 3-clique, by the union bound (Theorem 10.2), is at most the number of five-node graphs in which $\{1, 2, 3\}$ is a clique, plus the number of five-node graphs in which $\{1, 2, 4\}$ is a clique, plus the number of five-node graphs in which $\{1, 2, 5\}$ is a clique, plus ... plus the number of five-node graphs in which $\{3, 4, 5\}$ is a clique. This is a gross overcounting but it will be good enough. For general k and n, there are $\binom{n}{k}$ terms here.

By the way, all these $\binom{n}{k}$ numbers are the same! This is because there's nothing special about any one subset. Hence the number of n-node graphs containing a k-clique is at most $\binom{n}{k}$ times the number of n-node graphs containing a clique on $\{1, 2, 3, ..., k\}$.

To pick a random graph on n vertices, there are $\binom{n}{2}$ "coin-flipping" decisions we have to make: is there an edge between vertices 1 and 2; is there an edge between vertices 1 and 3; and ... and is there an edge between vertices $n - 1$ and n? There are $\binom{n}{2}$ "slots" for edges. Since there are $\binom{n}{2}$ decisions, there are $2^{\binom{n}{2}}$ graphs on n vertices.

How many of the $2^{\binom{n}{2}}$ n-node graphs have a clique on $\{1, 2, ..., k\}$? Well, if we know $\{1, 2, ..., k\}$ defines a clique, then those $\binom{k}{2}$ edges all definitely exist. In this case, we only have to make $\binom{n}{2} - \binom{k}{2}$ coin-flipping decisions. Hence the number of n-node graphs containing a clique on $\{1, 2, ..., k\}$ is $2^{\binom{n}{2} - \binom{k}{2}}$.

Now we apply the union bound (Theorem 10.2), summing over the potential locations of the k-clique. Since there are $\binom{n}{k}$ possible locations of a clique, and all the counts are the same, the number of n-node graphs having a k-clique *anywhere* is at most

$$\binom{n}{k} 2^{\binom{n}{2} - \binom{k}{2}}.$$

Using the fact that the complement operation switches cliques and independent sets, we immediately infer that the number of n-node graphs having an independent set of size k anywhere is the same and hence also at most

$$\binom{n}{k} 2^{\binom{n}{2} - \binom{k}{2}}.$$

Unfortunately, now we have to do some calculations. The conclusion of the calculations will be that

$$\frac{\binom{n}{k} 2^{\binom{n}{2} - \binom{k}{2}}}{2^{\binom{n}{2}}} \le \frac{1}{6},$$

that is, the probability of choosing a random n-vertex graph having a k-clique is at most $1/6$, or equivalently,

$$\binom{n}{k}2^{\binom{n}{2}-\binom{k}{2}} \leq \frac{1}{6}\cdot 2^{\binom{n}{2}},$$

that is, at most one-sixth of the n-node graphs contain a k-clique. Similarly, the number of n-node graphs containing an independent set of size k is at most $1/6$ of the total number of n-node graphs. Put them together and what do you conclude? That the number of n-node graphs containing a k-node clique or an independent set of size k is at most $1/3$ of the total number of graphs! This means at least one graph is left over. That left-over graph has no clique of size k and no independent set of size k. That's what we were trying to prove.

The only thing left is the calculations. Here we go. Remembering that $\binom{k}{2} = k(k-1)/2$ and canceling the $2^{\binom{n}{2}}$ from the numerator and denominator, we need to show that

$$\binom{n}{k}2^{-(k^2-k)/2} \leq \frac{1}{6}.$$

Since $\binom{n}{k} \leq n^k/k!$ and $n = \lfloor 2^{k/2}\rfloor$, it certainly suffices to prove that

$$\frac{\lfloor 2^{k/2}\rfloor^k}{k!}2^{-(k^2-k)/2} \leq \frac{1}{6},$$

and hence it also suffices to prove that

$$\frac{(2^{k/2})^k}{k!}2^{-(k^2-k)/2} \leq \frac{1}{6}.$$

But $(2^{k/2})^k = 2^{k^2/2}$, which cancels with the $2^{-k^2/2}$. Hence it suffices to prove that

$$\frac{2^{k/2}}{k!} \leq \frac{1}{6}.$$

We assumed $k \geq 4$. At $k = 4$, we want to prove that $2^{4/2}/24 \leq 1/6$, both sides of which are $1/6$. For a general $k \geq 4$, we have $2^{k/2} = (\sqrt{2})^k$ so

$$\frac{2^{k/2}}{k!} \leq \left(\frac{1}{6}\right)\frac{(\sqrt{2})^{k-4}}{5\cdot 6\cdot 7\cdots k} \leq \left(\frac{1}{6}\right)\left(\frac{\sqrt{2}}{5}\right)^{k-4} \leq \frac{1}{6},$$

and we are done.   ∎

It was a lot of work, no? But what a beautiful proof!

10.7   The Central Limit Theorem

The *Central Limit Theorem* is truly a mathematical gem. It asserts, roughly, that there is one single family of probability distributions to which sums of independent samples from *any* distribution converge. This is an astonishing fact. Allow me to explain. I will simplify things to make them easier to understand. I will leave out some boring technical details.

A *(discrete) probability distribution* is an assignment of nonnegative reals, called *probabilities*, to a finite set of points, subject to the condition that the sum of the assigned probabilities is 1. For example, in the case of rolling a normal die, one would assign $1/6$ to each of the six outcomes 1, 2, 3, 4, 5, 6; the sum of $1/6$ over six integers is 1. In the case of birthdays (skipping leap days again), one would assign $1/365$ to each of the 365 days of the year, represented by 1, 2, 3, ..., 365. So far, we've looked at *uniform* distributions in which all the probabilities are the same, but there's no reason they need be the same. Consider flipping a *biased* coin, one in which heads (which we'll model as 0) is twice as likely as tails (which we'll model as 1). For this experiment, we should assign $2/3$ to 0 and $1/3$ to 1. You could imagine a strange die in which $P[1] = 0.3$, $P[2] = 0.1$, $P[3] = 0.4$, $P[4] = 0.1$, $P[5] = 0.06$, and $P[6] = 0.04$. The only requirement is that the probabilities be nonnegative and sum to 1.

There are also *continuous* probability distributions on infinite sets of reals. For example, one can draw a random real number between 0 and 1, all such reals being equally likely. Mostly, we won't get into continuous distributions because they're more complicated. Most of the time, we'll stick with discrete distributions.

Let's think about what happens when one rolls a "die" repeatedly. To keep the arithmetic easy, we'll roll a (strange) five-sided die in which each of the five "sides," labeled 1, 2, 3, 4, and 5, has a $1/5 = 0.20$ chance of landing on the top. (One can imagine doing this experiment without figuring out how to design such a die.) Let's draw a bar chart showing the probability of each outcome. Look at Fig. 10.1. It shows five outcomes, 1, 2, 3, 4, and 5, each with probability $1/5 = 0.2$. No surprise here.

What happens if we roll the strange die twice and add up the two numbers we get? Let's write an outcome as a pair (i, j), which means that the first "die" landed with i on top and the second "die" landed with j on top. Here are the possible outcomes:

- The smallest sum we can get is 2, which we can only get by $(1, 1)$ (there is one way to get 2);
- the next smallest is 3, which we can only get as $(1, 2)$ and $(2, 1)$ (there are two ways to get 3);
- then 4, which can only be obtained as $(1, 3)$, $(2, 2)$, $(3, 1)$ (three ways to get 4);
- then 5, which can only be obtained as $(1, 4)$, $(2, 3)$, $(3, 2)$, $(4, 1)$ (four ways to get 5);

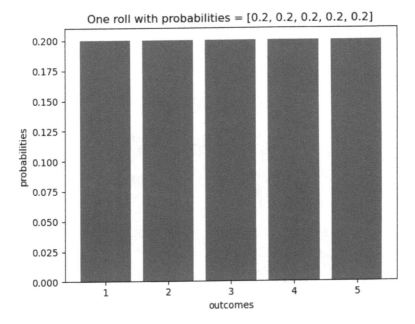

Fig. 10.1 Rolling a uniform five-sided "die" once

- then 6, which can only be obtained as $(1, 5), (2, 4), (3, 3), (4, 2), (5, 1)$ (five ways to get 6);
- then 7, which can only be obtained as $(2, 5), (3, 4), (4, 3), (5, 2)$ (only four ways to get 7);
- then 8, which can only be obtained as $(3, 5), (4, 4), (5, 3)$ (only three ways to get 8);
- then 9, which can only be obtained as $(4, 5), (5, 4)$ (only two ways to get 9);
- and last, 10, which can only be obtained as $(5, 5)$ (only one way to get 10).

Each particular pair, like $(4, 5)$ or $(4, 2)$, has probability exactly $0.2 \cdot 0.2 = 0.04$ of occurring. Therefore the probability of an outcome in $\{2, 3, 4, 5, 6, 7, 8, 9, 10\}$ is just 0.04 times the number of outcomes. For example, the probability of getting a sum of 6 is $0.04 \cdot 5 = 0.2$. See Fig. 10.2 for the bar chart.

Now let's roll three times and sum the three numbers we get. See Fig. 10.3 for the bar chart. The smallest possible sum is 3, obtainable only from $(1, 1, 1)$, whose probability is $0.2 \cdot 0.2 \cdot 0.2 = 0.008$. The next smallest smallest sum is 4, obtainable from $(2, 1, 1), (1, 2, 1)$, and $(1, 1, 2)$. Each of the three outcomes has probability 0.008, so the probability of getting a sum of 4 is $3 \cdot 0.008 = 0.024$. The next smallest sum is 5, obtainable from $(3, 1, 1), (1, 3, 1), (1, 1, 3), (1, 2, 2), (2, 1, 2)$, and $(2, 2, 1)$. Since there are six outcomes with sum 5, each of which has probability 0.008, the probability of getting a sum of 5 is $6 \cdot 0.008 = 0.048$.

Let's skip to ten rolls now. (It's fortunate that we have computers to do these calculations.) See Fig. 10.4.

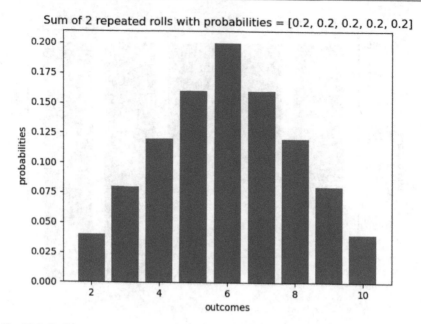

Fig. 10.2 Rolling a uniform five-sided "die" twice

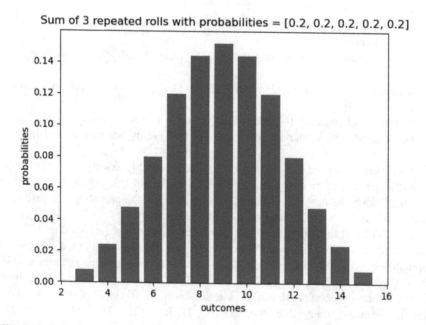

Fig. 10.3 Rolling a uniform five-sided "die" three times

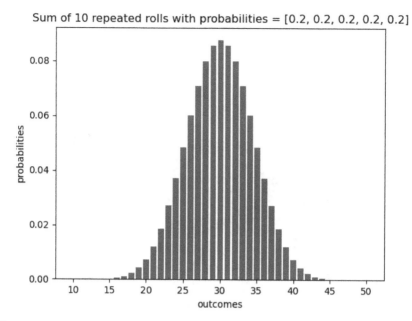

Fig. 10.4 Rolling a uniform five-sided "die" ten times

Notice the shape. It resembles that of a bell; it's been not-so-creatively named a "bell-shaped curve." By the way, notice the labels on the x-axis. If you roll a 5-sided die 10 times, the smallest possible sum is 10 and the largest possible sum is 50.

If you were to roll more than 10 times, you'd get a bar chart that resembles a bell even more.

You're probably looking at the bar chart and thinking, "Wow, that's a cool chart. There's probably nothing special about the fact that there are five outcomes; probably something similar would happen if there were six or seven outcomes instead. The picture must be so nice because all the probabilities are the same." OK. Let's try doing the same thing for probabilities 0.4, 0.2, 0.1, 0.05, 0.25 (which add up to 1, as they must). First, we roll once. See Fig. 10.5. Not surprisingly, we get the given probabilities.

Now we roll twice and add the two numbers. See Fig. 10.6. Still kind of ugly.

How about three rolls? See Fig. 10.7. Still kind of ugly.

Maybe it's looking a bit better to you. Let's also try four rolls; see Fig. 10.8. Magically, it's starting to look like a bell.

See Fig. 10.9 for ten rolls.

Voila! Almost bell-shaped again. You probably don't think there's anything special about the fact that there are five outcomes, and there's certainly nothing special about the probabilities we used. It seems natural to conjecture that whatever probability distribution you start with, if you take more and more independent samples from the distribution, and add the outcomes, you'll get a bell-shaped

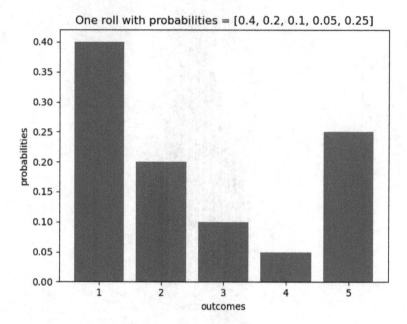

Fig. 10.5 Rolling a nonuniform five-sided "die" once

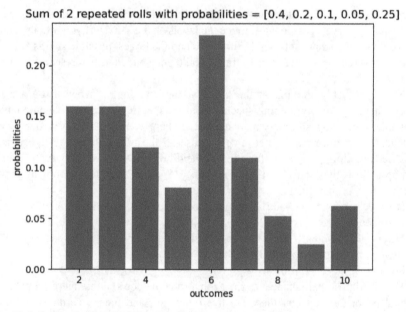

Fig. 10.6 Rolling a nonuniform five-sided "die" twice

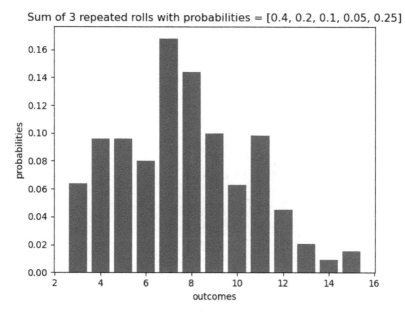

Fig. 10.7 Rolling a nonuniform five-sided "die" three times

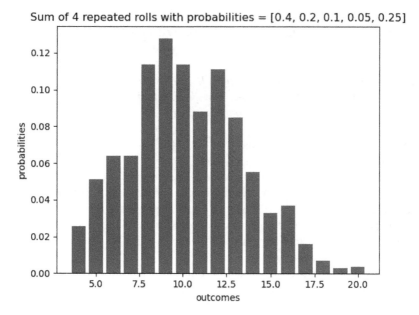

Fig. 10.8 Rolling a nonuniform five-sided "die" four times

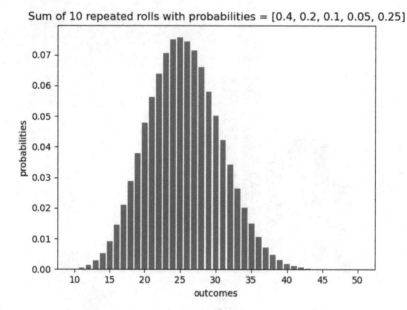

Fig. 10.9 Rolling a nonuniform five-sided "die" ten times

curve, and you'd be correct.[4] This is the Central Limit Theorem, which I will state somewhat imprecisely (after all, I haven't formally defined "bell-shaped" and I have omitted the technical condition).

Theorem 10.4 *Central Limit Theorem 1. If you take n independent samples from any distribution and add the outcomes, then the distribution of the sum will approach a bell-shaped distribution as n approaches infinity.*

It is really a remarkable fact that "one size fits all"—whatever distribution you started with, you end up with the same shape.

In fact, there's more. If you adjust the labels on the *x*-axis appropriately (by subtracting the mean and then dividing by the standard deviation, if you know what those words mean), there is only *one* bell-shaped curve that you will always approach, regardless of the starting distribution. This is a unique distribution, and is independent of the distribution you started with. The distribution is so special, it's called the *normal distribution*.

But wait! There's more, once again. It turns out that the *n* independent samples can be chosen from *different* distributions, subject to a technical condition which I won't describe. This is even more remarkable than the Central Limit Theorem given in Theorem 10.4.

---

[4] In fact, you need a technical condition that the distribution have finite variance, but that condition almost always holds, so I won't bother defining variance here.

Theorem 10.5 *Central Limit Theorem 2. If you take independent samples from n possibly different distributions, the ith sample from the ith distribution, and add the outcomes, then (subject to a condition which I'm omitting) the distribution of the sum will approach a bell-shaped distribution as n approaches infinity.*

There is even a version of the Central Limit Theorem that doesn't require full independence.

The Central Limit Theorem is hard to prove, so we won't prove it. Unfortunately the formula for the normal distribution is complicated. The precise formula of the "bell" is

$$y = \frac{1}{\sqrt{2\pi}} e^{-x^2/2},$$

where $\pi = 3.14159...$ and $e = 2.718...$ was introduced in Chap. 1. Figure 10.10 is a careful plot of the true bell-shaped curve.

Why is the Central Limit Theorem so important? The Central Limit Theorem says that whenever you add up (almost) independent random quantities, you approach a normal distribution. If something, like a person's height, is the sum of many factors, if they are not too dependent on each other, the final result will be approximately normally distributed. For example, men's heights are approximately normally distributed. So are women's heights. If you go out and measure something in the real world, there's a good chance it will be approximately normally distributed. That's why, after all, it's called the "normal" distribution—it's what normally happens!

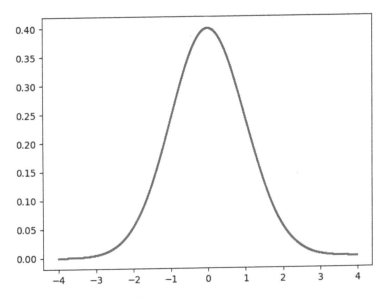

Fig. 10.10 The normal distribution

10.8   Puzzle

Suppose you have an $n \times n$ square S of chocolate where n is a positive integer. You want to cut S up into n^2 1×1 squares. S is marked with lines at all integral distances from the square's edges. In one step, you can pick up *one* piece and make one cut, from one end to the other, on a marked line.

What is the minimum number of cuts necessary? Make sure you prove that it is impossible to make fewer cuts than the number you give.

10.9   Exercises

Exercise 10.1 A family has two children, at least one of which is a girl. What's the probability that both are girls?

Exercise 10.2 * You have a *biased* coin C, meaning that the probability of getting a head is $p \neq 1/2$. However, you don't know p. Show how to use C to generate an unbiased 0-1 bit.

Exercise 10.3 * Suppose you have a coin which has a probability p of landing heads. Show that the expected number of coin flips until one gets a head (including the flip landing heads) is $1/p$.

Exercise 10.4 * Suppose you are collecting coupons. (Trust me, you don't have to actually collect coupons.) There are n distinct coupons $C_1, C_2, C_3, ..., C_n$. There are an infinite supply of copies of C_1, an infinite supply of copies of C_2, an infinite supply of copies of C_3,..., and an infinite supply of copies of C_n.

When you pick a coupon, you get uniformly at random, one of $\{C_1, C_2, C_3, ..., C_n\}$. You stop as soon as you've collected at least one of each coupon. Show that the average number of coupons you will have to draw

$$= \left(1 + \frac{n}{n-1} + \frac{n}{n-2} + \frac{n}{n-3} + \cdots + \frac{n}{1} \right)$$

$$= n \left(\frac{1}{n} + \frac{1}{n-1} + \frac{1}{n-2} + \frac{1}{n-3} + \cdots + \frac{1}{1} \right).$$

Exercise 10.5
(a) Use calculus to calculate the optimal number of tickets to buy in the casino night problem, as a function of n, the number of tickets already sold.
(b) When $n = 180$, calculate the probability of winning and the expected profit, if one buys the optimal number of tickets.

Exercise 10.6 Here is a statistics problem that occurs often in the real world. You have an enormous but finite universe of points, each labeled "positive" or "negative."

The positives are bad; you wish they didn't exist. Let $p = 0.05$. You want to show that the fraction f of positives in the universe is less than p, but the universe is too large to examine each point.

In such a scenario, it is natural to *sample* from the universe, that is, to pick a random sample S of n points from U, and examine the points in S. The experiment is very simple.

1. If the number of positives in S is zero, print "The fraction of positives in the universe is less than p" (which may or may not be true).
2. Otherwise, print "I don't know."

The problem is that with n small, like 1, the experiment is likely to print "The fraction of positives in the universe is less than p" even when it is not true. Even if n is large, like 1000, there is always some positive probability of making a mistake. When the fraction of positives in the universe is p or greater, there is still a positive probability that the sample S consists only of negatives (so that the experiment will erroneously output "The fraction of positives in the universe is less than p").

Suppose we are given a *confidence* target c of 99%. This just means that we want the probability that the experiment erroneously outputs "The fraction of positives in the universe is less than p," when it's not true, to be at most $1 - c = 0.01$.

Assume that the fraction f of positives in the universe is exactly $p = 0.05$ so that "The fraction of positives in the universe is less than p" is false. Because the universe is so large, assume that every time a point from the universe is chosen for the sample, it has a probability $p = 0.05$ of being positive, and that the choices are independent. Find the least n such that the probability that all n points in S are negatives (and hence the experiment outputs an erroneous answer) is at most $1 - c = 0.01$. With this size sample, if the experiment outputs "The fraction of positives in the universe is less than p," we can be 99% confident that the statement is true.

Now do the same for confidence $c = 0.95$.

Exercise 10.7 Drawing a point at random from a 2×2 square, what's the probability that it lands in the unit disk centered at the center of the square?

Exercise 10.8 Here is an example of a *Markov chain*. A system is always in state 1 or state 2.

If it is in state 1, it stays in state 1 with probability $1/4$ and moves to state 2 with probability $3/4$.

If it is in state 2, it stays in state 2 with probability $1/3$ and moves to state 1 with probability $2/3$.

It starts in state 1.

(a) Show that after one step, it is in state 1 with probability $1/4$ and in state 2 with probability $3/4$.
(b) What are the probabilities of being in states 1 and 2 after two steps?

(c) What are the probabilities of being in states 1 and 2 after three steps?

Exercise 10.9 * Suppose one flips an unbiased coin (one with probability of heads equaling $1/2$) repeatedly. On average, how many flips must one make until one gets two consecutive heads?

Exercise 10.10 * You are given *oracle access* to a polynomial f with real coefficients. This means that you can specify one value of x, say, $x = a$, and ask for, and receive, the value $f(a)$. Assume there is no rounding error so that $f(a)$ is exact. You do not get to see the coefficients of the polynomial.

The polynomial f may be the zero polynomial. You are given, however, a positive integer d and promised that *if $f(x)$ is nonzero*, then its degree is at most d.

You are given a small positive number ϵ. Design an algorithm that makes one query and returns either "f is the zero polynomial" or "f is not the zero polynomial" with the following property:

- if f is the zero polynomial, then your algorithm returns "f is the zero polynomial," and
- if f is not the zero polynomial, then your algorithm incorrectly returns "f is the zero polynomial" at most a fraction ϵ of the time (returning the correct answer "f is not the zero polynomial" the remainder of the time).

Assume your algorithm has a supply of random numbers.

Exercise 10.11 You are given a random variable X which takes value x_1 with probability p_1, value x_2 with probability p_2, value x_3 with probability p_3, ..., and value x_n with probability p_n. You want to devise a "summary statistic" s for your distribution, that is, one number that summarizes the distribution. Specifically, you want s to minimize

$$\sum_{i=1}^{n} p_i (x_i - s)^2.$$

Show that you should choose s to be the mean $\sum_i p_i x_i$ of X. Warning: this problem requires calculus.

Exercise 10.12 * Given a sorted sequence $< x_1, x_2, x_3, \ldots, x_n >$ with $x_1 \leq x_2 \leq \cdots \leq x_n$, the *median* is defined to be the middle value $x_{(n+1)/2}$ if n is odd, and instead the average of the two middle values if n is even. Hence the median of $< 3, 5, 6, 7, 10 >$ is 6.

In this problem, to keep things simple, we will assume that the x's are all different and that n is odd.

You are given a sorted sequence $< x_1, x_2, x_3, \ldots, x_n >$ with $x_1 < x_2 < \cdots < x_n$ and n odd. You want to devise a "summary statistic" t for your distribution, that is, one number that summarizes the sequence. Specifically, you want t to minimize

$$f(t) = \sum_{i=1}^{n} |x_i - t|.$$

Show that $f(t)$ is minimized when t is the median of the sequence.

Exercise 10.13 I am 68 inches tall. Heights of American men are approximately normally distributed with mean value of 69 inches and standard deviation of 2.5 inches. Using the normal distribution approximation, estimate what fraction of American men are shorter than I am.

The *standard* normal distribution, which is the continuous probability distribution shown in Fig. 10.10, has mean 0 and standard deviation 1. The curve shown in Fig. 10.10 is the *density function* for the distribution. For example, the chance that a random point generated is near 0 is about 0.40. More precisely, given a small positive ϵ, the chance that the random point generated is between $-\epsilon$ and ϵ is about $0.40 \cdot (2\epsilon)$. Similarly, the chance that the random point generated is between $-2 - \epsilon$ and $-2 + \epsilon$ is about $0.05 \cdot (2\epsilon)$. One has to jump through these annoying hoops with the ϵ because the normal distribution is continuous.

That the mean is 0 means that the center of the curve is at 0; in particular, half of the mass of the distribution will be negative and half positive. (This is informal and is not the definition of "mean.") The standard deviation is a measure of how wide the curve is. Given arbitrary μ and positive number σ (pronounced "sigma"), one can get a normal distribution having mean μ and standard deviation σ by taking the standard normal distribution, multiplying it by σ and then adding μ to it. To be more precise, if X is a random number generated from the standard normal distribution of mean 0 and standard deviation 1, then $X' = \sigma X + \mu$ is a normally distributed random variable of mean μ and standard deviation σ.

To get the fraction of points in the *standard* normal distribution of value at most a given t, one can use this Python program:

```
from scipy.stats import norm
t = 0
print("CDF(",t,") =", norm.cdf(t))
```

cdf standing for "cumulative distribution function," and changing the value of t. Notice that for $t = 0$, you get 0.5, as you should. Note that you might have to run "pip install scipy" before running Python.

Your job is to figure out how to use this Python code to estimate the fraction of American men at most 68 inches tall.

Exercise 10.14 In a discrete probability space, there is a finite set $W = \{w_1, w_2, ..., w_n\}$, the ith point w_i having a nonnegative probability p_i, the p_i's summing to 1. A *random variable* X is a mapping from W to the reals, that is, X assigns a real number to each point w_i. One usually omits the argument when referring to X. One writes things like, "$P[X = 3] = 0.4$," which means that the sum of the p_i's for which $X(w_i) = 3$ is 0.4. The *expected value $E[X]$ of X* is defined to be the sum over i of $X(w_i) \cdot p_i$. This is its "mean" or "average" value. It follows immediately from the definition that if X and Y are two random variables on W and $Z = X + Y$, then $E[Z] = E[X] + E[Y]$.

Consider any random variable X. Let μ be its mean value. Obviously two different random variables can have the same mean. For example, one random variable can be 0 with probability 1, and the other can be 1000000 half the time and -1000000 half the time; these are very different random variables, both of mean 0. It's often important to get some idea of how much a random variable typically differs from its mean. Let's call this, informally, the "spread" of the random variable. How should one define spread? There are different definitions, but the one which is easiest to use is the "variance." It uses the *square* of the distance from the mean.

Definition 10.2 The *variance* var(X) of a random variable X of mean μ is $E[Y]$ where $Y = (X - \mu)^2$. The *standard deviation* is the square root of the variance.

For example, if a random variable X is defined on $W = \{w_1, w_2, w_3\}$ with $X(w_1) = 1$, $X(w_2) = 3$, and $X(w_3) = 8$ with all three points having probability $p_i = 1/3$, then $\mu = E[X] = 1 \cdot (1/3) + 3 \cdot (1/3) + 8 \cdot (1/3) = (1 + 3 + 8)/3 = 4$. This means $Y(w_1) = (1 - 4)^2 = 9$, $Y(w_2) = (3 - 4)^2 = 1$, $Y(w_3) = (8 - 4)^2 = 16$. This makes the variance $9 \cdot (1/3) + 1 \cdot (1/3) + 16 \cdot (1/3) = 26/3$.

(a) Prove that var$(X) = E[X^2] - (E[X])^2$.

This formula confused me when I first saw it. The first term is the expected value of the random variable X^2. The second term is the square of the expected value of X. These are not the same! In the example, $E[X^2] = (1/3)(1^2 + 3^2 + 8^2) = 74/3$ and $E[X] = 4$, so $E[X^2] - (E[X])^2 = 74/3 - 16 = 74/3 - 48/3 = 26/3$.

(b) Prove that for any X, $E[X^2] \geq (E[X])^2$.

(c) Show that if var$(X) = 0$, then there is a number c such that the probability that $X = c$ is 1.

(d) We know that $E[X + Y] = E[X] + E[Y]$. Show that var$(X + Y)$ may *not* equal var$(X) +$ var(Y).

(e) Analogous definitions and theorems to those above apply to continuous random variables. If you let X be the standard normal distribution (see Exercise 10.13), which has mean 0 and standard deviation 1, what is $E[X^2]$?

Exercise 10.15 Let $1 \leq k \leq n - 1$. Prove that

$$\binom{n}{k} = \binom{n-1}{k} + \binom{n-1}{k-1}$$

without doing any calculations at all. No factorials!

10.9.1 Hints

10.2 Flip the biased coin twice. Use some of the four possible outcomes and discard others.

10.3 Let x be the expected value. Write an equation for x in terms of itself and solve for x. Specifically, the expected number of flips is p (the probability that the first flip is a head) times 1 (the number of coin flips that were necessary in this case to get a head), plus $1 - p$ (the probability that the first flip is a tail) times the expected number of coin flips if the first flip is a tail (which is how much?).

10.4 Suppose you already have collected i distinct coupons and you are seeking your $(i + 1)$st. Calculate the probability that the next coupon you collect is a new one and use Exercise 10.3.

10.9 The answer is neither 4 nor 8. See the exercise on Markov chains. Build a Markov chain with states "Start," "Just flipped a head," and "Done," with appropriate transition probabilities. Write down and solve equations for the average number of flips necessary to reach the "Done" state from each state.

10.10 Use a fact about the number of roots of a nonzero polynomial of degree at most d.

10.12 Show that if t is not the median, then one can slightly increase or decrease t to decrease $f(t)$. This is called a "local improvement" argument.

Fractals

11

11.1 The Cantor Set and the Sierpiński Triangle

This book is about mathematical beauty. Usually, when I think of beauty in mathematics, I'm thinking of elegance and simplicity. However, in mathematics there is also beauty of the artistic kind, probably the best example of which is that of fractals. Informally, fractals are sets that reveal more and more detail at smaller and smaller scales. No matter how deeply you look, you find interesting structure.

Let's start off with a very simple example, the Cantor set, named after the same German mathematician Georg Cantor who came up with the diagonalization argument to prove that the set of reals is uncountable (see Chap. 6). The notation (a, b), where a, b are real numbers with $a < b$, means the set of all real numbers between a and b, excluding a and b. This is known as an *open interval*. For example, the open interval $(1/3, 2/3)$ means all reals between $1/3$ and $2/3$, except for $1/3$ and $2/3$.

Here's a definition of the Cantor set:

1. Let $I_0 = [0, 1]$, the *closed interval* of all reals between 0 and 1, including both endpoints.
2. Remove the open middle third $(1/3, 2/3)$ from I_0, and call the new set I_1. I_1 is a disjoint union of the two intervals $[0, 1/3]$ and $[2/3, 1]$ (and contains all four endpoints).
3. Remove the middle third (as an open interval) from each of the two intervals of I_1 and call the new set I_2. I_2 is a disjoint union of the four intervals $[0, 1/9]$, $[2/9, 3/9]$, $[6/9, 7/9]$, $[8/9, 1]$.
4. Remove the middle third (as an open interval) from each of the four intervals of I_2 and call the new set I_3. I_3 is a disjoint union of the eight intervals $[0, 1/27]$, $[2/27, 3/27]$, $[6/27, 7/27]$, $[8/27, 9/27]$, $[18/27, 19/27]$, $[20/27, 21/27]$, $[24/27, 25/27]$, and $[26/27, 1]$.
5. Continue in this way forever.

Fig. 11.1 The Cantor set, in which each point is shown as a vertical line

The Cantor set C is the final set of points, those that remain after middle thirds are repeatedly removed forever.

There are infinitely many points in the Cantor set, so I cannot draw all of them perfectly on a screen with a finite number of pixels, but see Fig. 11.1 for an approximate picture, in which each point in the Cantor set is represented as a vertical red line.

It is also possible to describe the Cantor set noniteratively. Instead of using a decimal representation to represent reals, we can use *base-3* or *ternary representation*, using digits 0,1,2. This is just like decimal representation, but uses powers of 3 instead of powers of 10. For example, the number $210.210022_3 = 2 \cdot 9 + 1 \cdot 3 + 0 \cdot 1 + 2/3 + 1/9 + 0/27 + 2/81 + 2/243$. Using this ternary representation, the Cantor set C is the set of all reals between 0 and 1, including 0 and 1, which have an infinite ternary representation having no ones (i.e., having only 0's and 2's). (Note that $1/3$ is in the Cantor set because $1/3 = 0.0222222...._3$). This equivalence is not obvious and should really be proven but I won't bother.

What makes the Cantor set so interesting is that it is self-similar. The left half of the Cantor set (the part in $[0, 1/3]$) is a copy of the full Cantor set, shrunken by a factor of 3; likewise for the right half of the Cantor set (the part in $[2/3, 1]$).

What is the dimension of the Cantor set? We haven't discussed dimension yet, but you probably know that an interval has dimension 1, that a square has dimension 2, that a cube has dimension 3, and so on. Since C is a subset of $[0, 1]$, you'd probably have guessed that its dimension would be at most 1... but would you have guessed that its dimension was $(\log 2)/(\log 3)$, which is approximately 0.6309? (Whenever you see a logarithm without a base in an expression, the base of the log is irrelevant. You can choose any base larger than 1 (e.g., 2, 10, or $e = 2.718...$) that you want, as long as you use the same base throughout that expression.)

How can a dimension not be an integer? What does "dimension" mean anyway? We will see later.

Now that we've seen one instance of a fractal, let's see another. Here's one of them, the Sierpiński triangle, in Fig. 11.2. This figure is a subset of $\mathbb{R}^2$, unlike the Cantor set (but its dimension is not 2). It is obtained from the equilateral triangle by removing the (open) interior of the middle (upside-down) triangle, and then removing the interior of the middle upside-down triangle in each of the remaining triangles, and then removing the interior of the middle upside-down triangle in each of the remaining triangles, and so on, forever. Its dimension, as we will see later, is $(\log 3)/(\log 2)$, which is approximately 1.585.

Fig. 11.2 The Sierpiński triangle

11.2   The Mandelbrot Set

11.2.1 Background

Here, in Fig. 11.3, is a beautiful and famous fractal, known as the *Mandelbrot set*, which is named after Benoit Mandelbrot, a pioneer in the study of fractals.

How is the Mandelbrot set defined? Remember the complex numbers we studied in Chap. 5? We need them in order to define the Mandelbrot set. Recall that a complex number z is a number of the form $x + yi$ where x and y are real numbers and $i = \sqrt{-1}$. You can plot a complex number $z = x + yi$ at the point (x, y) in the Euclidean plane, exactly as if $z = x + yi$ were the point (x, y). In Chap. 5 we saw how to add, subtract, multiply, and divide complex numbers.

Since we understand how to do arithmetic on complex numbers, we are almost ready to define the Mandelbrot set. For any complex number c, let $f_c(z)$ be the function that takes a complex number z and returns $z^2 + c$, which is another complex number. For example, if $c = 3 + i$, then $f_c(i) = i^2 + (3+i) = -1 + (3+i) = 2+i$.

The Mandelbrot set depends on an infinitely repeated application of the function $f_c(z)$, starting from 0. In other words, you pick a complex number c and write down the infinite sequence $< 0, f_c(0), f_c(f_c(0)), f_c(f_c(f_c(0))), \ldots >$. Of course, you can't really write down an infinite sequence, but you can imagine doing so. Now

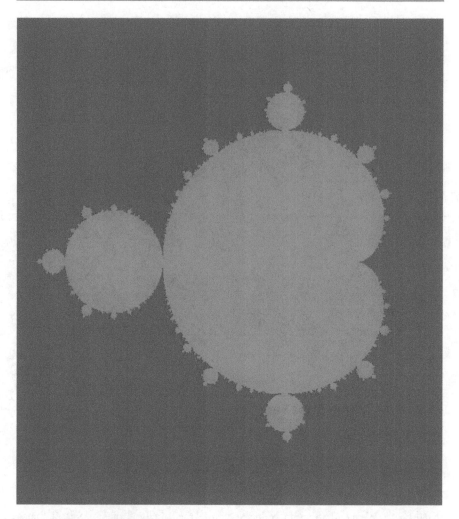

Fig. 11.3 The Mandelbrot set

there are two possibilities: (1) that the sequence "goes off to infinity," whatever that means, and (2) it does not.

The first possibility, which is called *diverging*, must be defined carefully. Recall from Chap. 5 that the "norm" or "length" of a complex number $z = x + yi$ is denoted $|z|$ and is defined to be $\sqrt{x^2 + y^2}$. Then we say an infinite sequence $< z_1, z_2, z_3, \ldots >$ of complex numbers *diverges* if for any positive real number R, there is a complex number in the sequence whose length is at least R. For example, this sequence diverges: $< 1, i, -1, -i, 2, 2i, -2, -2i, 3, 3i, -3, -3i, 4, 4i, -4, -4i, \ldots >$. There are four points of length 1; then four points of length 2; then four points of length 3; then four points of length 4; etc. For any positive R, there is a point in this sequence of length at least R. We can think of this sequence as "going off to infinity."

By contrast, here is a sequence that doesn't diverge: $< 17, 17i, -17, -17i,$ $17, 17i, -17, -17i, 17, 17i, -17, -17i, ... >$. This series does not diverge, since no number in it has length exceeding 17, but do notice that while it is not diverging, it is not approaching any particular complex number, either. It does not "converge."

11.2.2 Definition of the Mandelbrot Set

Now that we have the background, we can define the Mandelbrot set.

Definition 11.1. The *Mandelbrot set* is the set of all complex numbers c such that the infinite sequence $< 0, f_c(0), f_c(f_c(0)), f_c(f_c(f_c(0))), ... >$ does not diverge.

For example, take $c = 1 = 1 + 0i$, so $f_c(z) = z^2 + 1$. The infinite sequence mentioned is $< 0, 1, 2, 5, 26, 677, 458,330, ... >$, which clearly diverges, so $c = 1$ is *not* in the Mandelbrot set.

Try $c = i$. We get the infinite sequence beginning $< 0, 0^2 + i = i, i^2 + i = -1+i, (-1+i)^2 +i = -i, (-i)^2 +i = -1+i >$. Since the $-1+i$ appears twice, the sequence will now repeat forever. Hence it does not diverge and hence i *is* in the Mandelbrot set.

In practice, when you're drawing the Mandelbrot set, (1) you can't look at *all* c's, because there are infinitely many; (2) you can't write down the whole infinite sequence $< 0, f_c(0), f_c(f_c(0)), f_c(f_c(f_c(0))), ... >$; and (3) you can't even compute $f_c(z)$ exactly, because there will be roundoff error. You are forced to make some approximations. There are two sources of error, other than roundoff error, and we will ignore roundoff error. A point is in the Mandelbrot set if the infinite sequence $< 0, f_c(0), f_c(f_c(0)), f_c(f_c(f_c(0))), ... >$ does not diverge. Instead of looking at *divergence* of an *infinite* sequence, instead we'll ask if the first, say, 100, points have length at most, say, 10.

Python makes working with complex numbers really easy. You define complex number $z = x + yi$ by saying $z = complex(x, y)$. To get the absolute value of z, you just execute $abs(z)$. For example, this program

```
z = complex(3, 4)
print("The absolute value of", z, "is", abs(z))
```

prints "5" for the absolute value of z, as you will see if you run it yourself. You will also discover that Python uses j, not i, for the square root of -1. Nonetheless, I will continue to use i for the square root of -1.

Now, we're finally ready to explain how to draw the Mandelbrot set. Instead of asking if the infinite sequence $< 0, f_c(0), f_c(f_c(0)), f_c(f_c(f_c(0))), ... >$ does not diverge, we'll only ask if the first *num_iterations* iterations lead to z's of absolute value at most *threshold*. So here's some code that determines whether a point c is (approximately) in the Mandelbrot set. We will color points in the Mandelbrot set red (which is $(255, 0, 0)$ in RGB space) and those not in the Mandelbrot set blue $((0, 0, 255)$ in RGB space):

```python
from PIL import Image, ImageDraw

def mandelbrot(c):
  num_iterations = 100
  threshold = 10
  z = complex(0, 0)
  for n in range(num_iterations):
    z = z * z + c
    if abs(z) > threshold:
      # Not in Mandelbrot set so blue.
      return (0, 0, 255)
  # In Mandelbrot set so red.
  return (255, 0, 0)
```

11.2.3 Drawing the Mandelbrot Set

Now that we can figure out whether a point is (approximately) in the Mandelbrot set or not, we need a program to actually draw the Mandelbrot set. Here's the code (which you can find on the book's website, so that you don't need to type it all in). It uses the mandelbrot_module.py program. You can draw the Mandelbrot set yourself:

```python
from PIL import Image, ImageDraw
from mandelbrot_module import mandelbrot

if __name__ == "__main__":

  # The plotting window goes from smallest_real to
  # largest_real on the x-axis and from
  # smallest_imaginary
  # to largest_imaginary on the y-axis.
  smallest_real = -1.5
  largest_real = 0.68
  smallest_imaginary = -1.2
  largest_imaginary = 1.2

  # Break the x-range into scale_factor pieces.
  scale_factor = 300
  width_in_pixels = int(scale_factor * \
    (largest_real - smallest_real))
  # Do the same for the y-range.
```

```
height_in_pixels = int(scale_factor * \
  (largest_imaginary - smallest_imaginary))

# (0, 0, 0) means black background
image = Image.new('RGB', (width_in_pixels, \
  height_in_pixels), (0, 0, 0))
draw = ImageDraw.Draw(image)

# Let x vary from 0 to width_in_pixels - 1.
for x in range(width_in_pixels):
  # Let y vary from 0 to height_in_pixels - 1.
  for y in range(height_in_pixels):
    # Convert pixel coordinates to a complex number
    # with real part between smallest_real
    # and largest_real and with imaginary part
    # between smallest_imaginary
    # and largest_imaginary.
    real_part = smallest_real + \
      (x / width_in_pixels) * \
      (largest_real - smallest_real)
    imaginary_part = smallest_imaginary + \
      (y / height_in_pixels) * \
      (largest_imaginary - smallest_imaginary)
    c = complex(real_part, imaginary_part)
    color_vector = mandelbrot(c)
    # Plot the point (x, y) with the given
    # color_vector.
    draw.point([x, y], color_vector)

image.save('mandelbrot_output.png', 'PNG')
```

This program will create an output file called "mandelbrot_output.png." On a Windows computer, at least, you should be able to view the image by double-clicking on it.

11.3   The Newton Fractal

This fractal should be of particular interest to anyone, like you, who has just studied Newton's method. It turns out that, when there are multiple roots to an equation, determining to *which* root Newton's method will converge, when started at a given starting point, is complicated. In this case, we are seeking the roots to the equation $z^3 - 1 = 0$ over the complexes, whose roots are $z = 1$, $z = (-1 + i\sqrt{3})/2$, and $z = (-1 - i\sqrt{3})/2$. The Python program juliadraw3.py colors points that converge to the first point red; those that converge to the second point, green; and

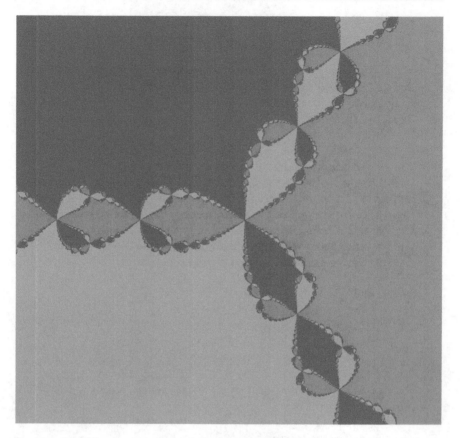

Fig. 11.4 The Julia set

those that converge to the third point, blue. (The complex number $z = x + yi$ is plotted at (x, y).) What you get is the fractal known as a *Julia set* and which is shown in Fig. 11.4.

The code to generate the image is `juliadraw3.py`, shown here, and also on the website:

```
from PIL import Image, ImageDraw
import math

def f(z):
  # f(z) = z^3 - 1.
  return z * z * z - 1.0

def fprime(z):
  # The slope (or derivative) of
  # f(z) = z^3 - 1 is 3 z^2.
```

```
    return 3.0 * z * z

def julia(z, num_iterations, tol):
  # The three complex cube roots of 1.
  roots = [complex(1.0, 0.0), \
    complex(-0.5, math.sqrt(3.0) / 2.0), \
    complex(-0.5, -math.sqrt(3.0) / 2.0)]
  # red, green, blue
  colors = [(255, 0, 0), (0, 255, 0), (0, 0, 255)]

  for i in range(num_iterations):
    # Check to see if z is close to a root.
    # If so, return the color of that root.
    for j in range(len(roots)):
      if abs(z - roots[j]) <= tol: return colors[j]
    # No color found, so do a step of
    # Newton iteration.
    deriv = fprime(z)
    if deriv == 0:
      return (0, 0, 0) # black
    # Newton iteration.
    z = z - f(z) / deriv

  # If haven't converged yet, color the point black.
  return (0, 0, 0)

if __name__ == "__main__":
  # The plotting window goes from smallest_real to
  # largest_real on the x-axis and from
  # smallest_imaginary to largest_imaginary
  # on the y-axis.
  smallest_real = -1.75
  largest_real = 1.5
  smallest_imaginary = -1.5
  largest_imaginary = 1.5

  scale_factor = 300
  width_in_pixels = int(scale_factor * \
    (largest_real - smallest_real))
  height_in_pixels = int(scale_factor * \
    (largest_imaginary - smallest_imaginary))

  image = Image.new('RGB', (width_in_pixels, \
    height_in_pixels), (0, 0, 0))
  draw = ImageDraw.Draw(image)
```

```
# Let ix vary from 0 to width_in_pixels - 1.
for ix in range(width_in_pixels):
  # Let iy vary from 0 to height_in_pixels - 1.
  for iy in range(height_in_pixels):
    # Convert pixel coordinates to a complex
    # number with real part x between
    # smallest_real and largest_real and with
    # imaginary part y between
    # smallest_imaginary and largest_imaginary.
    x = smallest_real + (ix / width_in_pixels) * \
      (largest_real - smallest_real)
    y = smallest_imaginary + \
      (iy / height_in_pixels) * \
      (largest_imaginary - smallest_imaginary)
    z = complex(x, y)
    color_vector = julia(z, 100, 0.000001)

    # Plot the point (ix, iy) with the given
    # color_vector.
    draw.point([ix, iy], color_vector)

  image.save('julia_set_output3.png', 'PNG')
```

11.4   Box-Counting Dimension

11.4.1 Introduction

As mentioned above, we all know that the dimension of a point is 0, the dimension of a line segment is 1, the dimension of a square is 2, the dimension of a cube is 3, and so on. But why? And how should one define the dimension of a fractal like the Cantor set or the Sierpiński triangle? We will see shortly that the dimension of a fractal may be *nonintegral*. In particular, the dimension of the Cantor set is $(\log 2)/(\log 3)$, which is approximately 0.6309. This seems odd.

The key is to define "dimension" correctly. There are several different definitions, which are related to each other but not identical. I will show you the simplest one, the "box-counting dimension." It's very concrete. Let's look at a set of points in $\mathbb{R}^d$. For example, the Cantor set C is a subset of $\mathbb{R}^1$.

I assume you know what a d-dimensional *cube* is, so I won't define it. I mean cubes that are not "tilted": their sides are parallel to the coordinate axes. A one-dimensional cube is a line segment; a two-dimensional cube is a square; and a three-dimensional cube is called, well, a cube. Cubes of dimension four or higher are sometimes called *hypercubes*. We are going to be counting cubes. However, in the case of box-counting dimension, we will sometimes call them *boxes*.

Here is the main question: given a subset S of $\mathbb{R}^d$, how many d-dimensional cubes of side length ϵ do I need to "cover" (i.e., contain) S, and how quickly does that number $n_\epsilon(S)$ grow as ϵ shrinks toward 0?

For example, let's take $d = 2$ and a 2×3 rectangle $R = \{(x_1, x_2) | 0 \le x_1 \le 2, 0 \le x_2 \le 3\}$. Take any positive and small ϵ. What is the minimum number of $\epsilon \times \epsilon$ cubes I need in order to cover R? If $1/\epsilon$ is an integer, then I can divide the first axis into $2 \cdot 1/\epsilon$ parts and the second axis into $3 \cdot 1/\epsilon$ parts. We can draw $(2/\epsilon) \cdot (3/\epsilon) = 6/\epsilon^2$ cubes (squares) the union of which covers R. The key point here is that the (reciprocal of the) number of cubes needed looks like ϵ *to the second power*; hence the box-counting dimension will be 2. This is why the dimension of a rectangle is 2. We will see later that the definition of box-counting dimension ensures that the 6 disappears. In general, the length and width of the rectangle don't matter at all. The dimension of every rectangle which has length a and width b, provided only that both a and b are positive, is two.

Needless to say, if you used $\mathbb{R}^3$ and studied a set like $\{(x_1, x_2, x_3) | 0 \le x_1 \le 1, 0 \le x_2 \le 2, 0 \le x_3 \le 3\}$ in $\mathbb{R}^3$, you'd see that the number of cubes necessary (when $1/\epsilon$ is an integer) would be $(1/\epsilon)(2/\epsilon)(3/\epsilon) = 6/\epsilon^3$, and the box-counting dimension would be 3.

Now the (minor) challenge remaining is to give a formal definition that works for most sets. Intuitively, the box-counting dimension is the exponent b such that the number of cubes you need looks like $1/\epsilon^b$, at least when ϵ is very small. That is, we want to define b so that $n_\epsilon(S)$ equals $(1/\epsilon)^b$. If they were exactly equal, for every ϵ, we'd have $\log n_\epsilon(S) = \log((1/\epsilon)^b) = b \log(1/\epsilon) = -b \log \epsilon$. That means b would equal $-\log n_\epsilon(S)/(\log \epsilon)$ for every ϵ. However, it's unreasonable to expect this to hold simultaneously for every ϵ. That means b should equal $-\log n_\epsilon(S)/(\log \epsilon)$... but for which ϵ? Since we only care about this formula for extremely small ϵ, we'll look at the smallest ϵ's, by defining the box-counting dimension of S to be $\lim_{\epsilon \to 0}[-\log n_\epsilon(S)/\log \epsilon]$.

If you've never seen a limit before, don't panic. The limit is just what the expression becomes as ϵ gets closer and closer to 0. All the uses of "limit" in this chapter will be very easy.

Definition 11.2 (Box-Counting Dimension, Definition 1). If the minimum number of cubes of side length ϵ we need to cover S is $n_\epsilon(S)$, then the *box-counting dimension* of S is defined to be the limit, as ϵ approaches 0, of $-\log n_\epsilon(S)/\log \epsilon$.

(By the way, the limit may not exist, but we won't discuss this possibility further.)

Sometimes it is more convenient when computing dimension, not to look at all positive ϵ's, but instead to look at boxes whose side lengths are powers of, say, $1/2$ (like $1/2, 1/4, 1/8$, etc.), or of, say, $1/3$ (like $1/3, 1/9, 1/27$, etc.). There is nothing special, of course, about $1/2$ and $1/3$. One need look only at boxes whose side lengths are powers of c for any fixed c, $0 < c < 1$. Let's give another definition.

Definition 11.3 (Box-Counting Dimension, Definition 2). Let $0 < c < 1$. For k a positive integer, the minimum number of cubes of side length c^k we need to cover S was already defined as $n_{c^k}(S)$. The *box-counting dimension* of S is defined to be the limit (if it exists), as k approaches infinity, of $-\log n_{c^k}(S)/\log c^k$.

The interesting and not obvious fact is that both definitions of box-counting dimension give the same result (and therefore changing the value of c won't change box-counting dimension, either). While not too hard, the proof is technical and not so interesting so I will not give it. But the equivalence between the definitions means we can use either definition we want (and any value of c we want, provided it's strictly between 0 and 1, if we're using the second definition), when calculating box-counting dimension.

This definition gives box-counting dimension 1 for a line segment, 2 for a square, and 3 for a (3-dimensional) cube. The interesting question is, what does it give for sets like the Cantor set or the Sierpiński triangle?

11.4.2 The Cantor Set

Computing how many boxes of a given size are needed to cover a set may sound difficult, but in some cases, it's easy. It's easy for the Cantor set.

Let's compute the box-counting dimension of the Cantor set. Remember that the left half of the Cantor set is a copy of the full Cantor set, except that it's scaled by a factor of $1/3$; the same is true of the right half of the Cantor set. Because we scale by a factor of $1/3$, we will use the second definition of box-counting dimension, and we will use $c = 1/3$. Let c_k be the number of boxes of side length $(1/3)^k$ needed to cover the Cantor set. Then the dimension is the limit, as k goes to infinity, of $(\log c_k)/(\log 3^k) = (\log c_k)/(k \log 3)$.

Let's start by computing c_0. We need to find the minimum number of (closed) intervals of length $(1/3)^0 = 1$ needed to cover C. Given that the Cantor set is a subset of $[0, 1]$, the one interval $[0, 1]$ will cover C. Furthermore, you clearly need at least one interval of side length 1—zero intervals clearly won't work—so $c_0 = 1$.

Now let's compute c_1. We need to compute the minimum number of closed intervals of side length $(1/3)^1 = 1/3$ needed to cover C. Clearly the two intervals $[0, 1/3]$ and $[2/3, 1]$ suffice, so $c_1 \leq 2$, but do we know that $c_1 = 2$? Are we sure that one interval of length $1/3$ cannot work? In fact, it cannot work; we will prove this later. The point I'm making is that there is something that needs to be proven.

Now let's compute c_2. What is the minimum number of closed intervals of side length $(1/3)^2 = 1/9$ needed to cover C? By the construction of C, these four intervals suffice: $[0, 1/9]$, $[2/9, 3/9]$, $[6/9, 7/9]$, $[8/9, 1]$. Hence $c_2 \leq 4$, but do we know that no three strange intervals of side length $1/9$ can cover C? These three intervals could have irrational endpoints, as does $[1/\pi, 1/\pi + 1/9]$. How can we be sure that no three strange intervals of side length $1/9$ can cover C? There is something to be proven here.

In general, it is easy to see that 2^i intervals of side length $(1/3)^i$ cover C. However, is this the *minimum* number? It is, as we will soon see. Let's just assume it for now.

Theorem 11.1. *Assuming* $c_i = 2^i$, *the dimension of the Cantor set is* $(\log 2)/(\log 3)$.

Proof. We saw above that the dimension of any set is the limit, as k goes to infinity, of $(\log c_k)/(\log 3^k)$ (because we are covering with boxes of side length $(1/3)^k$). This expression equals $(\log c_k)/(k \log 3)$. Assuming that $c_k = 2^k$, we need the limit of $(\log 2^k)/(k \log 3) = (k \log 2)/(k \log 3) = (\log 2)/(\log 3)$, whose limit of course is $(\log 2)/(\log 3)$. ■

Now let's patch the hole we left. We know that $c_i \leq 2^i$, but how do we know that $c_i = 2^i$? In light of the fact that $c_i \leq 2^i$, proving $c_i \geq 2^i$ will prove that $c_i = 2^i$.

Lemma 11.1. $c_i \geq 2^i$.

Proof. In constructing C, we start with $[0, 1]$ and then repeatedly remove the open middle thirds. In the ith step, 2^i intervals remain, each of length $(1/3)^i$. The key point is that the *left endpoints* of those intervals are in C. (That is, they are not removed in future steps.) For example, I_2 is a disjoint union of the four intervals $[0, 1/9], [2/9, 3/9], [6/9, 7/9], [8/9, 1]$ with left endpoints $0, 2/9, 6/9, 8/9$, all of which are in C. I_3 is a disjoint union of the eight intervals $[0, 1/27], [2/27, 3/27], [6/27, 7/27], [8/27, 9/27], [18/27, 19/27], [20/27, 21/27], [24/27, 25/27]$, and $[26/27, 1]$, with left endpoints $0, 2/27, 6/27, 8/27, 18/27, 20/27, 24/27, 26/27$, all of which are in C. The important observation here is that in I_i, two different left endpoints differ by at least $2 \cdot (1/3)^i$, so that no interval of length $(1/3)^i$ could possibly contain two of the left endpoints. Since those 2^i left endpoints are in C, we need at least 2^i intervals of length 3^{-i} to cover C. ■

We have proven that the box-counting dimension of the Cantor set is $(\log 2)/(\log 3)$.

11.4.3 The Sierpiński Triangle

In this section we will compute the dimension of the Sierpiński triangle, which we'll call S. We saw above that when computing box-counting dimension, we didn't need to compute the minimum number of boxes of side length ϵ necessary, for all positive ϵ. It was enough to look at boxes of side length 2^{-k} or 3^{-k}. In fact, you don't even need boxes! You can use other simple shapes, like equilateral triangles one of whose sides is horizontal—let's call these *horizontal equilateral triangles*—when working in dimension 2. Informally, the reason is that a horizontal equilateral triangle of side length ϵ can be covered by a fixed number (which is independent of ϵ) of slightly smaller boxes, and a box of side length ϵ can be covered by a fixed number (which is independent of ϵ) of smaller horizontal equilateral triangles. The bottom line is that it suffices to determine the minimum number s_i of equilateral triangles (with one side horizontal) of side length 2^{-i} one needs to cover S. Then the box-counting dimension of S is $\lim_{i\to\infty}[-\log s_i / \log(2^{-i})] = \lim(-\log s_i)/(-i\log 2) = \lim(\log s_i)/(i\log 2)$.

Remember how S was constructed. It was a lot like how the Cantor set was constructed, except the action occurred in $\mathbb{R}^2$ instead of $\mathbb{R}^1$. We started with a horizontal equilateral triangle of side length 1 and removed the interior of the "upside-down" equilateral triangle of side length $1/2$ from the center, leaving 3^1 horizontal equilateral triangles each of side length $1/2^1$. From each of the three side-length-$1/2$ horizontal equilateral triangles, we removed the interior of an upside-down equilateral triangle of side length $1/4$. This left us with 3^2 horizontal equilateral triangles, each of side length 2^{-2}. Then, after removing the interior of the middle equilateral triangle of each, we have 3^3 horizontal equilateral triangles, each of side length 2^{-3}. In general, in the ith round, there are 3^i horizontal equilateral triangles, each of side length 2^{-i}. This set of triangles covers S, so it follows that $s_i \leq 3^i$. We are interested in

$$\lim_{i\to\infty} [-\log s_i / \log 2^{-i}] = \lim(\log s_i /(i\log 2)).$$

We know that $s_i \leq 3^i$, so the limit, if it exists, will be at most

$$\lim[\log 3^i /(i\log 2)] = \lim[(i\log 3)/(i\log 2)] = (\log 3)/(\log 2).$$

Now we are left with a dilemma similar to the one we had for the Cantor set. Can we prove that $s_i \geq 3^i$? Or can we prove that the box-counting dimension of S is at least $(\log 3)/(\log 2)$ without proving that $s_i \geq 3^i$?

It turns out that we have to be only slightly more clever than we were in the proof of Lemma 11.1. Pretty easily one can prove that $s_i \geq 3^{i-1}$ and that will be enough to prove that the dimension of S is $(\log 3)/(\log 2)$.

Lemma 11.2. $s_i \geq 3^{i-1}$.

Proof. Consider the 3^i horizontal equilateral triangles of side length 2^{-i} we got in the ith round. Look at the top corner of each triangle. These top corners are distinct—the top corner of one triangle is never the top corner of another—so there are 3^i top corners. All of these points are in S (i.e., they are never removed). Any set of triangles of side length 2^{-i} that covers S also covers the 3^i top corners. Let S_i be the set of top corners. The key point is that S_i is a subset of S for every i, so that $s_i = n_{2^{-i}}(S) \geq n_{2^{-i}}(S_i)$.

Fact 11.1 *No single side-length-2^{-i} horizontal equilateral triangle can contain four points of S_i.*

Proving this fact is left as Exercise 11.1.

It follows that in order to cover the 3^i top corners, one needs at least $3^i/3 = 3^{i-1}$ side-length-2^{-i} horizontal equilateral triangles. ∎

Now we just finish up.

Theorem 11.2. *The box-counting dimension of S is $(\log 3)/(\log 2)$.*

Proof. $\lim_{i\to\infty}[-\log s_i/\log 2^{-i}] = \lim[\log s_i/(i\log 2)]$. We have $3^{i-1} \leq s_i \leq 3^i$, so $(i-1)\log 3 \leq \log s_i \leq i\log 3$. Hence $(i-1)\log 3/(i\log 2) \leq \log s_i/(i\log 2) \leq (i\log 3)/(i\log 2) = (\log 3)/(\log 2)$. Now we have $(\log s_i)/(i\log 2)$ sandwiched between $[(i-1)/i][(\log 3)/(\log 2)]$ at the bottom and $(\log 3)/(\log 2)$ at the top. Clearly $(\log 3)/(\log 2)$ approaches $(\log 3)/(\log 2)$ as i goes to infinity. The key point is that as i goes to ∞, $[(i-1)/i]$ approaches one, so $[(i-1)/i][(\log 3)/(\log 2)]$ approaches $(\log 3)/(\log 2)$ as well. That proves that $(\log s_i)/(i\log 2)$ approaches $(\log 3)/(\log 2)$. ∎

11.5   Puzzle

This puzzle is attributed to Martin Gardner, who, for 25 years, wrote a "Mathematical Games" column in the magazine *Scientific American*, and whom I had the honor of meeting once at a "Gathering for Gardner" celebration in Atlanta.

A man lives in Manhattan, New York, near the 34th Street and 7th Ave. subway station. In this unusual station, there is a central platform serving both uptown and downtown express trains. The man has two girlfriends, one lives uptown and one, downtown. Every Saturday evening the man arrives at a random time, stands on the center subway platform, and awaits an express train to take him to visit his girlfriend. But which girlfriend does he visit? The uptown girlfriend, if the uptown train comes first, or the downtown girlfriend, if that train arrives first.

After a year of visiting girlfriends, he checks his records and discovers that he has visited the uptown girlfriend on 90% of the Saturdays and the downtown girlfriend on 10% of the Saturdays.

How can this be? Uptown and downtown trains both arrive every 10 minutes, and he arrives on the platform at a random time. Shouldn't he have visited both girlfriends about the same number of times? And why is the downtown girlfriend still seeing him after a year of infrequent visits?

11.6   Exercises

Exercise 11.1. * Prove Fact 11.1.

Exercise 11.2. The *TCP fractal* is one of my personal favorites, since it was defined in a paper I coauthored with mathematician Anna Gilbert when we were both working at ATT Labs—Research [3]. It comes with a nice backstory.

TCP stands for Transmission Control Protocol, which is usually used when computers send data across the Internet. The basic question is, how fast should a source computer send data? Data is sent in packets. If the source sends so much data so quickly that the recipient can't read it all, some of the data will be lost and the recipient will have to ask for a retransmission, wasting time. If there's enough congestion on the edges of the Internet that data packets are lost and never get to the recipient, the source will have to resend the data. On the other hand, if the edges are uncongested and the recipient is idle, then the source should send data at as high a rate as feasible. Why not use the available bandwidth if no one else is using it?

To maximize throughput, TCP uses an "additive increase, multiplicative decrease" protocol. Specifically, each source maintains a "rate" at which it will send data. If it sends a packet successfully, it slightly increases its rate, additively, in fact (this is the "additive increase" part). If the packet is not received successfully, the source "backs off" by dramatically decreasing its rate, by a factor of 2 (this is the "multiplicative decrease" part).

Here is the precise model Anna and I proposed. There are two parameters, a *buffer size B* and a *drain rate d*. Two sources 0 and 1 maintain rates r_0 and r_1, respectively, starting with some positive values. A recipient has one buffer of size $B \geq d$ into which both sources compete to inject packets. The buffer's occupancy— that is, how full it is—is represented by b ($0 \leq b \leq B$) and the buffer is initially full. In each time step:

1. The recipient drains d units from its buffer, but of course never lets the buffer occupancy b go negative: Set $b = \max\{b - d, 0\}$.
2. An $i \in \{0, 1\}$ is selected uniformly at random.
3. Source i "fires," which means that first r_i is increased by 1 (an additive increase), and then the buffer occupancy b is increased by r_i, except, of course, that it can never exceed B: Set $b = \min\{b + r_i, B\}$.
   Source i now backs off if the buffer is currently full (since that's a sign of data loss), that is, if $b = B$, source i sets $r_i = r_i/2$ (a multiplicative decrease).
4. Now the other source $j = 1 - i$ fires: r_j is increased by 1 (an additive increase), and b is set to $\min\{b + r_j, B\}$.
   If $b = B$, source j sets r_j to $r_j/2$ (a multiplicative decrease).

This process is repeated ad infinitum.

Run this code, also available on the website, to draw TCP fractals. The code generates file "output.png", which you can then view:

```
import random
from PIL import Image, ImageDraw

def fire(i, b, r, B):
  # Additive increase
  r[i] += 1.0
  # Send data.
  b = min(b + r[i], B)
  # If the buffer is full, source i "backs off"
  # (multiplicative decrease).
  if b == B:
    r[i] /= 2.0
  return b, r

def run_one_step(b, r, B, d):
  # The buffer is partially drained.
  b = max(b - d, 0)

  # Randomly decide who fires first.
  rand = random.random()
  i = 0 if rand < 0.5 else 1

  # Source i in {0, 1} fires first.
  b, r = fire(i, b, r, B)
  # Source 1 - i fires next.
  b, r = fire(1 - i, b, r, B)
  return b, r

if __name__ == "__main__":

  # A good value for B is 4.0.
  B = float(input("Enter buffer size B: "))
  # Start with a full buffer.
  b = B
  # A good value for d is 3.0.
  d = float(input("Enter the drain rate d: "))
  r = [1.0, 1.0] # initial rates
  # A good value for num_time_steps is 1000000.
  num_time_steps = int(input( \
    "Enter the number of time steps: "))

  min_x = min_y = 1.0
  max_x = max_y = d

  width_in_pixels = height_in_pixels = 1000
```

```
# (255, 255, 255) means white background.
image = Image.new('RGB', \
   (width_in_pixels, height_in_pixels), \
   (255, 255, 255))
draw = ImageDraw.Draw(image)

for step in range(num_time_steps):

   b, r = run_one_step(b, r, B, d);

   # These are the x plot upper bound
   # and the y plot upper bound.
   xplotub = 1000; yplotub = 1000
   if min_x <= r[0] <= xplotub and \
     min_y <= r[1] <= yplotub:
     transformed_r0 = (r[0] - min_x) * \
       (width_in_pixels) / (max_x - min_x)
     # Without the "height_in_pixels -" part,
     # small values of r[1] would be near the top.
     transformed_r1 = \
       height_in_pixels - (r[1] - min_y) * \
       (height_in_pixels) / (max_y - min_y)
     # (0, 0, 0) means black.
     draw.point([transformed_r0, \
       transformed_r1], (0, 0, 0))

   image.save('tcp_output.png', 'PNG')
```

(a) Draw the TCP fractal for $B = 40/7, d = 20/7$.
(b) Draw the TCP fractal for $B = 3000, d = 51$.

Exercise 11.3. Draw the $B = 4, d = 3$ TCP fractal using `tcp.py`. Imagine that, in the image that it generates, the x- and y-coordinates range from 0 to 1. Consider the upper-left corner H of the image (in which x goes from 0 to 1/2 and y goes from 1/2 to 1). Somewhere in the image there appears to be a $1/4 \times 1/4$ scaled-down copy of H. Where is it?

Exercise 11.4. Define the *tent map* $t(x)$ on $[0, 1]$ by $t(x) = (3/2)(1 - |2x - 1|)$, so called because plotted, it looks like a tent.

(a) Plot the tent map.
(b) Let C be the Cantor set.
    Show that $t(C) \subseteq C$, that is, if $x \in C$, then $t(x) \in C$.

(c) Show that $t(C) \supseteq C$, that is, for any $y \in C$, there is an $x \in C$ such that $t(x) = y$.

You have shown that $t(C) = C$.

Exercise 11.5. Some sets T, like the Cantor set and the Sierpiński triangle, are *strictly self-similar*, meaning that T is composed *exactly* of disjoint pieces each of which is a scaled-down and possibly rotated copy of itself. For such a set T, define its *similarity dimension* to be $-\log m / \log r$ if there are m pieces each being a copy of T scaled by r, $0 < r < 1$. The drawback to similarity dimension, as compared to, say, box-counting dimension, is that it is only defined on strictly self-similar sets.

(a) Show that the similarity dimension of the Cantor set equals its box-counting dimension.
(b) Show that the similarity dimension of the Sierpiński triangle equals its box-counting dimension.

Exercise 11.6. We're going to draw the *von Koch snowflake curve*.

Draw a line segment F_0 of length 1 starting at $(0, 0)$ and ending at $(1, 0)$ in the Euclidean plane.

Now for $i = 0, 1, 2, 3, \ldots$, do the following.

F_i consists of 4^i line segments, all of the same length l_i, joined "at an angle" between segments. To get F_{i+1} from F_i, walk along F_i from start to end. For each of the 4^i segments I of length l_i, break I into three segments each of length $l_i/3$; add to the left of the middle one B two segments L and R in such a way that B, L, and R form an equilateral triangle of side length $l_i/3$ ($B = base$, $L = left$, $R = right$); and then remove B from the picture.

For example, F_1 consists, in order, of a horizontal segment from $(0, 0)$ to $(1/3, 0)$ of length $1/3$; a diagonally up segment from $(1/3, 0)$ to $(1/2, \sqrt{3}/6)$ of length $1/3$; a diagonally down segment from that point to $(2/3, 0)$ of length $1/3$; and, last, a horizontal segment from $(2/3, 0)$ to $(1, 0)$ of length $1/3$. You can see F_0, F_1, and F_2 in Fig. 11.5.

If one takes larger and larger i's, the F_i's approach a limiting curve called the *von Koch snowflake curve* F. (One actually needs a proof that the F_i's do approach a limiting curve F.) See Fig. 11.6.

Show that the lengths of the F_i's approach infinity, as i goes to infinity.

Exercise 11.7. Compute the similarity dimension, which was defined in Exercise 11.5, of the von Koch snowflake, which was defined in Exercise 11.6.

Exercise 11.8. Convert `juliadraw3.py` into `juliadraw4.py` to build the Julia set for the equation $z^4 - 1 = 0$, and draw the fractal.

Fig. 11.5 Construction of
the von Koch snowflake

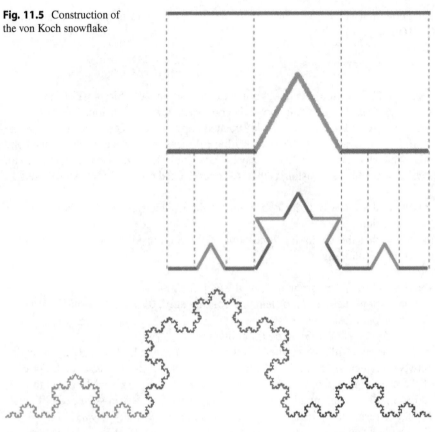

Fig. 11.6 The von Koch snowflake

11.6.1 Hint

11.1 Use (1) the fact that the maximum distance between any two points in a unit
equilateral triangle is 1, and (2) the fact that in a tesselation of the plane by
unit-side-length equilateral triangles, any set of four vertices contains two
points at distance more than one from each other.

Solutions to Puzzles 12

12.1 Chapter 1: Light Bulb Switches

Put all the switches in the off position and wait 45 minutes. After 45 minutes have elapsed, turn on switch 1, wait 14 minutes, and then turn it off. Then turn on switch 2. Immediately go into the next room and examine the three incandescent bulbs. If the bulb is on, it is controlled by switch 2. If it is off, feel the bulb. If it is cold, the bulb is controlled by switch 3, but if it is warm, then it is controlled by switch 1.

12.2 Chapter 2: Belt Around the Equator

Let r be the radius of the earth in meters. Then its equator has length $2\pi r$, as does the belt before one meter is added to it. Once a meter is added, the belt has length $2\pi r + 1$. Let $r' > r$ be the radius of the circle whose circumference is $2\pi r + 1$. Hence $2\pi r + 1 = 2\pi r'$ and hence $r' = r + 1/(2\pi)$. The key point is that $r' - r = 1/(2\pi)$, which is independent of r. Now $1/(2\pi)$ meters is about 16 centimeters or 6.26 inches. A bowling ball wouldn't fit but everything else would.

12.3 Chapter 3: Two Trains and a Fly

The temptation is to write down and sum an infinite geometric series, but this is the difficult way to solve the problem. Much easier is to think about time. The track is 100 kilometers long. The trains are approaching each other at 50 km/h, as T_1 travels at 20 km/h and T_2 at 30 km/h. This means the train will collide at 2:00 PM exactly. Flying at 40 km/h, the fly will have flown 80 kilometers before being crushed.

There's a famous anecdote about the great mathematician John von Neumann. When told the puzzle, he immediately replied with the correct answer. The person

H. Karloff, *Mathematical Thinking*, Compact Textbooks in Mathematics,
https://doi.org/10.1007/978-3-031-33203-6_12

telling him the puzzle replied, "I see you found the easy way, Professor von Neumann. Most people try to sum the infinite series."

"What easy way?", asked von Neumann. "I summed the series."

12.4   Chapter 4: Writing the Year as a Sum of Positive Integers with Maximum Product

Suppose that $n = a_1 + a_2 + \cdots + a_k$ with $a_1 a_2 \cdots a_k$ is as large as possible. (You get to choose k; it is not specified.)

The first observation is that none of the numbers can be five or larger, and in fact, there is a solution with no 4's. For if $x \geq 4$, we may replace x by the pair 2 and $x - 2$. It is easy to see that $2 \cdot (x - 2) > x$ if $x > 4$ and $2 \cdot (x - 2) = x$ if $x = 4$. This means that no optimal solution can contain a number at least 5, and if the optimal solution contains a 4, there's a different, also optimal, solution in which each 4 has been replaced by a pair of 2's.

Furthermore, there shouldn't be any 1's, since combining the 1 with another number would give a larger product. So now we know that there is an optimal solution with only 2's and 3's.

However, there cannot be even three 2's, for if there are three or more 2's, we can replace three 2's, which sum to 6 and have product 8, by two 3's, which sum to 6 also but have product 9. Hence there are zero, one, or two 2's, and the rest are 3's.

The only question is how many 2's there are. Since all the remaining numbers are 3's, $n \bmod 3$ will be 0 if there are no 2's, 2 if there is one 2, and 1 if there are two 2's (because $2 + 2 = 4$ has a remainder of 1 when divided by 3). So $n \bmod 3$ tells you how many 2's to take.

12.5   Chapter 5: Two Towers

There is a trick that avoids the square roots. Imagine inverting T_2 so that it runs from $(c, 0)$ to $(c, -b)$. The lengths of the two segments of chain remain the same. However, now it is obvious that to minimize the length of chain, the chain should run straight from $(0, a)$ to $(c, -b)$, which means that it should hit the x-axis at the point $(x, 0)$ where x satisfies

$$\frac{x}{a} = \frac{c - x}{b},$$

meaning (after a little algebra) that $x = \frac{a}{a+b} \cdot c$.

12.6   Chapter 6: A 10 × 10 Board

Color the 100 cells of the 10×10 board black and white as a chessboard is colored, horizontally or vertically adjacent cells getting opposite colors, the lower-left cell getting, say, black. Then the upper-right cell also gets black. There are 50 black cells and 50 white cells.

However, once the lower-left and upper-right cells are removed, 48 black cells and 50 white cells remain. If it were possible to cover the 98 cells with 49 dominoes, since each domino covers one black cell and one white cell, there would have to be 49 black and 49 white cells, so the task is impossible.

12.7   Chapter 7: Distribution of Genders

Every time a woman gives birth, there would still be a 50% chance of her giving birth to a boy and a 50% chance of her giving birth to a girl. There would be no change in the gender distribution.

12.8   Chapter 8: A Hunter and a Bear

The North Pole is one solution, but there are infinitely many other points besides the North Pole. Please try to find the remaining infinitely many other points before reading further.

Recall that lines of latitude run east-west around the earth. Just north of the South Pole, there is a line of latitude, which I will call C_1, whose length is one kilometer. Since on the scale of a few kilometers, the earth is approximately flat, the region near the South Pole (or anywhere else) is roughly Euclidean. This means that the line of latitude C_1 is approximately $1/(2\pi) \approx 0.159$ kilometers from the South Pole. If the hunter starts anywhere one kilometer *north* of C_1, walking south will take him or her to a point Z on C_1. Walking east around that line of latitude will return the hunter to Z, and then one kilometer north will take the hunter home. Already there are infinitely many solutions, since any point one kilometer north of C_1 returns the hunter home.

But wait, there are more! Let n be any positive integer. Define C_n to be a line of latitude so close to the South Pole that its length is $1/n$ kilometers. These lines of latitude are closer to the South Pole than C_1 (if $n > 1$). Walking one kilometer east on C_n will return a hunter, after walking along C_n n times, to the same point on C_n. Hence the hunter can start one kilometer north of any point on C_n.

These are the only solutions: the North Pole, together with the union over n, of all the points one kilometer north of any point on C_n.

Now let's prove these are the only solutions. Let's show that the only non-North Pole solutions are the ones I described. Close to the North Pole are lines of latitude C'_n, resembling C_n but being near the North Pole instead of the South Pole. However, all these lines of latitude are much less than one kilometer south of the North Pole, so the hunter couldn't start one kilometer north of any of them.

Now here's the key point: assuming the hunter starts at a point other than the North Pole, after walking east, he or she must be on the same line of longitude (not latitude) as his home. Otherwise there's no chance of ending at his or her home.

There are only two lines of latitude of length $1/n$ kilometers on the earth, and one is too close to the North Pole to help. The hunter's home must be (unless it's at the North Pole) one kilometer north of a point on C_n, so the stated points are the only solutions.

The bear was a white polar bear.   ∎

12.9   Chapter 9: A Mathematician's Children

The ages are three positive integers adding to 36. Here is a table of their possible ages, each list given in nondecreasing order, followed by the sum of the ages:

```
1, 1, 36: 38
1, 2, 18: 21
1, 3, 12: 16
1, 4,  9: 14
1, 6,  6: 13
2, 2,  9: 13
2, 3,  6: 11
3, 3,  4: 10
```

If the sum had been anything but 13, the census taker would have known the ages upon seeing the number of the house next door. This means that the number on the house next door must have been 13.

When told that the oldest child was at least a year older than all the others, the census taker knew the ages were 2, 2, 9.

12.10 Chapter 10: Cutting Up Chocolate

You can cut the bar of chocolate into n^2 1×1 pieces by making $n^2 - 1$ cuts, as follows. Until all the remaining pieces are 1×1, there must be a piece which is larger than 1×1. Pick up a piece which is larger than 1×1, choose any horizontal or vertical line inside the piece, and cut it along that line, cutting it into two pieces. Repeat, until all pieces are 1×1.

Let a_i be the number of pieces of chocolate remaining after i cuts have been made. Clearly $a_0 = 1$. Regardless of how one makes cuts, provided one cuts only one piece at a time, $a_{i+1} = a_i + 1$. By induction, this means that $a_i = i + 1$ for all i. Hence after $n^2 - 1$ cuts, n^2 pieces remain, regardless of how the pieces are cut, so all must be 1×1 (since every piece has area at least 1 and the total area is n^2). This means that no matter how you cut the chocolate, $n^2 - 1$ cuts suffice and are necessary.

12.11   Chapter 11: Uptown and Downtown Girlfriends

The answer is very simple. For example, the uptown trains can arrive at 0, 10, 20, 30, 40, and 50 minutes past the hour (e.g., at 6:00 PM, 6:10 PM, 6:20 PM, 6:30 PM, 6:40 PM, and 6:50 PM), while downtown trains arrive at 1, 11, 21, 31, 41, and 51 minutes past the hour. By arriving at a random time, 90% of the time the man arrives after a downtown train has left and before an arriving uptown train. Only 10% of the time does he arrive in the 1-minute gap after an uptown train and before a downtown one, for example, between 6:00 PM and 6:01 PM.

Interestingly, this puzzle has practical applications. Many years ago I lived in Hoboken, New Jersey, right across the Hudson River from Manhattan, for a year. Hoboken-to-New York buses were offered by two bus companies. The first one, New Jersey Transit, was a public agency. The second, Academy Bus, was a private company. Both companies ran their buses up Washington Ave, the main street in Hoboken, and they stopped at the same bus stops. When I waited for buses to New York on Washington Ave., it appeared that the Academy bus almost always came first. However, both the New Jersey Transit bus and the Academy bus ran with the same frequency. It occurred to me that Academy had set its schedule so that its buses would arrive just *before* the New Jersey Transit buses. I wonder what would have happened if New Jersey Transit had then altered its schedule to arrive just before Academy's buses, and Academy had then altered *its* schedule, and so on....

Acknowledgments

I thank Prof. Larry Riddle of Agnes Scott College for the two images pertaining to the von Koch snowflake, Aditya Makkar and Manish Chakrabarti for reviewing parts of the manuscript, and Koushik Balasubramanian for discussions about geometry and physics. I thank Prof. Steven Strogatz of Cornell University for discussions regarding writing and publishing a math book. I am grateful to Prof. John D. Norton of the University of Pittsburgh for discussions on proofs of the Pythagorean Theorem.

© The Author(s), under exclusive license to Springer Nature Switzerland AG 2023   191
H. Karloff, *Mathematical Thinking*, Compact Textbooks in Mathematics,
https://doi.org/10.1007/978-3-031-33203-6

References

1. J. A. Bondy and U. S. R. Murty, *Graph Theory With Applications*, North-Holland, New York, 1976.
2. Kenneth Falconer, *Fractal Geometry: Mathematical Foundations and Applications*, John Wiley & Sons, New York, 1990.
3. Anna C. Gilbert and Howard J. Karloff, "On the Fractal Behavior of TCP," *ACM Symposium on the Theory of Computing* (STOC), San Diego, 2003, 297–306.
4. Paul Erdős, as quoted in the article "Ramsey Theory" by Ronald L. Graham and Joel H. Spencer, in *Scientific American*, July, 1990, 112–117.

Index

Printed in the USA
CPSIA information can be obtained
at www.ICGtesting.com
LVHW011527270823
756422LV00001B/23